WARFARE AND TRACKING IN AFRICA, 1952–1990

Warfare, Society and Culture

Series Editors: David J. B. Trim
                   Andrew Wiest

Titles in this Series

1 Military Economics, Culture and Logistics in the Burma Campaign, 1942–1945
Graham Dunlop

2 Orde Wingate and the British Army, 1922–1944
Simon Anglim

3 The Jacobite Campaigns: The British State at War
Jonathan D. Oates

4 Arming the Royal Navy, 1793–1815: The Office of Ordnance and the State
Gareth Cole

5 Militant Protestantism and British Identity, 1603–1642
Jason White

6 The 1641 Depositions and the Irish Rebellion
Eamon Darcy, Annaleigh Margey and Elaine Murphy (eds)

7 Citizen Soldiers and the British Empire, 1837–1902
Ian F. W. Beckett (ed.)

8 Military Manpower, Armies and Warfare in South Asia
Kaushik Roy

9 Alexander Leslie and the Scottish Generals of the Thirty Years' War, 1618–1648
Steve Murdoch and Alexia Grosjean

10 German Soldiers in Colonial India
Chen Tzoref-Ashkenazi

Forthcoming Titles

War, Strategy amd the Modern State, 1792–1914
Carl Cavanagh Hodge

Worship, Civil War and Community, 1638–1660
Chris R. Langley

WARFARE AND TRACKING IN AFRICA, 1952–1990

BY

Timothy J. Stapleton

Routledge
Taylor & Francis Group

LONDON AND NEW YORK

First published 2015 by Pickering & Chatto (Publishers) Limited

2 Park Square, Milton Park, Abingdon, Oxfordshire OX14 4RN
52 Vanderbilt Avenue, New York, NY 10017

Routledge is an imprint of the Taylor & Francis Group, an informa business

First issued in paperback 2020

BRITISH LIBRARY CATALOGUING IN PUBLICATION DATA

Stapleton, Timothy J. (Timothy Joseph), 1967–, author.
Warfare and tracking in Africa, 1952–1990. – (Warfare, society and culture)
1. Tactics. 2. Tracking and trailing – Africa – History – 20th century. 3. Insurgency – Kenya – History – 20th century. 4. Insurgency – Zimbabwe – History
– 20th century. 5. Insurgency – Namibia – History – 20th century.
I. Title II. Series
355.4'22-dc23

ISBN-13: 978-1-8489-3558-7 (hbk)
ISBN-13: 978-0-367-59902-7 (pbk)

Typeset by Pickering & Chatto (Publishers) Limited

CONTENTS

Acknowledgements   vii
Abbreviations   ix

Introduction   1
1 Tracking and Identity   11
2 Tracking and Colonial Warfare   27
3 Kenya, 1952–6   43
4 Rhodesia (Zimbabwe), 1965–80   69
5 South West Africa (Namibia), 1966–90   101
Conclusion   137

Works Cited   145
Notes   161
Index   187

ACKNOWLEDGEMENTS

I would like to thank the Social Sciences and Humanities Research Council of Canada (SSHRC) for its financial support of this project. Furthermore, valuable assistance was rendered by the staff of the National Archives (Kew, United Kingdom), Imperial War Museum (London), the South African National Defence Force (SANDF) Documentation Centre, the National Archives and Library of Namibia, the National Archives of Botswana, the National Archives of Zambia, the William Cullen Library at the University of the Witwatersrand, the Stevenson-Hamilton Memorial Library in South Africa's Kruger National Park and Trent University's Inter-library Loan Department. Leon Bezuidenhout kindly arranged interviews in South Africa, and John Davis of the Kenya Regiment Association in Britain, Kevin Thomas in South Africa and Allan Savory in the United States provided otherwise unavailable documentation. Photographs were generously supplied by John Davis, Allan Savory and Jim Hooper.

ABBREVIATIONS

ANC	African National Congress
BDF	Botswana Defence Force
BSAC	British South Africa Company
BSAP	British South Africa Police
CMR	Cape Mounted Rifles
CAR	Central African Republic
CKG	Central Kalahari Game Reserve
CATU	Civilian African Tracking Unit, Rhodesia
DRC	Democratic Republic of Congo
DNPWLM	Department of National Parks and Wildlife Management, Rhodesia
FRELIMO	Front for the Liberation of Mozambique
FAMP	Frontier Armed and Mounted Police, Cape Colony
GAT	Guerrilla Anti-terrorist Unit, Rhodesia
JWS	Jungle Warfare School, British Army
KAU	Kenya African Union
KDF	Kenya Defence Force
KPR	Kenya Police Reserve
KCA	Kikuyu Central Association
KAR	King's African Rifles
LRA	Lord's Resistance Army
OZ	Oscar Zulu, Home Guard, South West Africa
MPLA	Movement for the Popular Liberation of Angola
FNLA	National Front for the Liberation of Angola
OAU	Organization of African Unity
PAC	Pan-Africanist Congress, South Africa
PLAN	People's Liberation Army of Namibia
PATU	Police Anti-terrorist Unit, Rhodesia
PRAW	Police Reserve Air Wing, Rhodesia
RR	Rhodesia Regiment
RAR	Rhodesian African Rifles
RLI	Rhodesian Light Infantry
RM	Romeo Mike Teams

ANC	African National Congress
RAF	Royal Air Force
SAAF	South African Air Force
SADF	South African Defence Force
SANDF	South African National Defence Force
SAP	South African Police
SWANU	South West African National Union
SWAPO	South West African People Union
SWASPES	South West African Specialist Unit
SWATF	South West African Territorial Force
SAS	Special Air Service
TCT	Tracker Combat Teams, Kenya
TCU	Tracker Combat Unit, Rhodesia
MK	Umkhonto we Sizwe
UDI	Unilateral Declaration of Independence
UN	United Nations
USAID	United States Agency for International Development
UDF	Union Defence Force
UNITA	Union for the Total Independence of Angola
VTU	Volunteer Tracking Unit, Rhodesia
ZANLA	Zimbabwe African Nationalist Liberation Army
ZANU	Zimbabwe African Nationalist Union
ZANU-PF	Zimbabwe African National Union – Patriotic Front
ZAPU	Zimbabwe African People's Union
ZNA	Zimbabwe National Army
ZIPRA	Zimbabwe People's Revolutionary Army
ZRP	Zimbabwe Republic Police

Contemporary Africa. Before 1990 Namibia was South West Africa, and before 1980
Zimbabwe was Rhodesia (Southern Rhodesia). Map reproduced courtesy of Daniel
Dalet, http://d-maps.com/carte.php?num_car=4339&lang=en.

During the Mau Mau Emergency in the 1950s, Kenya Game Department official Rodney Elliott (pictured), commissioned into the Kenya Regiment, became the founding commander of the British Army Tracking School at Nanyuki and aggressively advocated for the employment of Tracker Combat Teams. Photograph reproduced courtesy of John Davis, Kenya Regiment Association.

African trackers from the Kenya Regiment during the Mau Mau Emergency. Photograph reproduced courtesy of John Davis, Kenya Regiment Association.

In 1960 Allan Savory (pictured), an official of the Southern Rhodesia Game Department and a Territorial Army officer, was the first to suggest the training of military trackers by the Federation of the Rhodesias and Nyasaland. His ideas were slowly adopted by the federation's new Special Air Service (SAS), and in 1965 he formed the short-lived Guerrilla Anti-Terrorist Unit (GATU). During the late 1960s Savory founded the Rhodesian Army Tracker Combat Unit (TCU) comprised of white part-time soldiers who were hunters and game rangers in civilian life. GATU and TCU were forerunners of the Rhodesian Selous Scouts formed in the early 1970s. Photograph reproduced courtesy of Allan Savory.

In 1979 Johannes 'Sterk Hans' Dreyer (pictured centre) became the founding commanding officer of the South West African Police Counter-Insurgency Unit otherwise known as Koevoet. A career South African Police (SAP) officer, Dreyer had participated in counter-insurgency operations in Rhodesia and Mozambique during the early 1970s. Photograph reproduced courtesy of Jim Hooper.

In South West Africa during the 1980s, Koevoet Sergeant Shikongo Oholiko (pictured front), an expert in deciphering insurgent anti-tracking measures, leads the hunt. Photograph reproduced courtesy of Jim Hooper.

During the pursuit of South West African People Organization (SWAPO) insurgents, this Koevoet team has leap-frogged 1,000 to 1,500 metres ahead of the main tracking group to search for fresher signs. This tactic enabled Koevoet to gain time and distance on fleeing insurgents. Photograph reproduced courtesy of Jim Hooper.

Carrying a radio to keep the men in the Casspir armoured personnel carriers aware of the direction taken by fleeing SWAPO insurgents, Koevoet Sergeant Francois du Toit (pictured far left) joins his trackers on the spoor. With the Casspirs ready to provide cover fire in the event of ambush, the team's most experienced trackers use sticks to point out signs left by insurgents. Photographs reproduced courtesy of Jim Hooper.

In the wake of a firefight where three insurgents were killed, a South African Air Force Alouette gunship arrives to evacuate two badly wounded members of the Koevoet team. Photograph reproduced courtesy of Jim Hooper.

Ovambo soldiers from South West African Territorial Force (SWATF) 101 Battalion, a military unit inspired by Koevoet, run ahead of their Casspir vehicles tracking insurgents in northern South West Africa during the 1980s. Photograph reproduced courtesy of the South African National Defence Force Documentation Centre.

INTRODUCTION

Tracking is the ability to pursue and close with an animal or human subject by following signs, often called spoor which is an Afrikaans word, left behind in the environment. These signs include impressions in the ground such as footprints or scuff marks, disturbed vegetation, evidence of feeding, biological waste, sounds and smells. Experienced trackers determine the direction of their subject's movement by the appearance of these signs. Furthermore, the approximate age of spoor, which tells the tracker how far ahead in time and space the subject is, can be estimated by looking at how a sign has been changed over time by various factors such as drying of soil or broken foliage, movement of insects or animals that superimpose other signs, and weather especially rain. The position of the sun and resulting shadows are also important points in observing and interpreting spoor the conditions for which are best at first and last light. Tracking requires a thorough familiarity of the geography, climate and ecology of a specific locale as signs and the factors that affect them will be different in an open desert or a dense jungle. This skill also involves a great deal of informed speculation that allows a tracker who loses a trail to imagine the most likely path of the subject and attempt to pick up further signs in that direction. Catching up with prey is central to the exercise which means trackers usually avoid taking note of each and every sign directly in front of them, and prefer to look ahead to see spoor at a distance. When a knowledgeable tracker is being tracked, he or she can practise anti-tracking which means devising various methods to try to conceal signs or deceive the pursuer as to direction or age of spoor.

Among prehistoric human beings, tracking may have begun with the tracing of nocturnal animals to their daytime lairs. It eventually developed as a central feature of persistence hunting which involved the constant pursuit of prey over long distances. While humans run on two legs and reduce their body heat by sweating, four-legged animals lower their body temperature by slowing down to pant which meant early human hunters would eventually catch up to them. Another advantage possessed by early humans was that they could carry water in containers to maintain hydration during pursuits while animals could not. Persistence hunting was unique to early humans and played a major role in their physical and social evolution by promoting long-distance running on two legs and group cooperation. Arguably, within this process tracking had an impact on the growth of the human brain as natural selection favoured successful hunters with the ability to remember and interpret a large amount of complex spoor related information including a mental library of the tracks of different animal species. Perhaps more controversially, it has also been suggested that tra cking represented the beginning of early man's capacity to empathize as it involved predicting animal behaviour through imagining oneself in the place of the prey. This may have given rise to beliefs that expert hunters had the ability to spiritually enter the mind of the prey to determine where it was going. Furthermore, tracking promoted social formation as juveniles learned the skill by following, observing and copying more experienced adult hunters, and successful hunters might have attracted an entourage of dependent followers. Some scholars postulate that early man's tracking represented the origins of science as it required the collection and analysis of detailed data from the environment, and the formulation and testing of hypotheses. One study contends that tracking was first developed in dry and open environments where it was easier to follow footprints and that it spread to other areas relatively recently.[1] Critics of this view point to studies of modern hunter-gatherers that show persistence hunting conducted on the run is mostly practised in the open, flat and dry Kalahari Desert region of Southern Africa where tracking is supposedly easier and that those in savannah woodland environments like the Hadza of Tanzania usually abandon a hunt once the spoor is lost. Moreover, the majority of early humans occupied savannah-woodland environments where tracking is more challenging, it is difficult to run and track at the same time, and most modern studies of persistence hunting among surviving hunter-gatherers show this method to be rare and mostly unsuccessful. Therefore, it is hypothesized that most early humans pursued their prey by walking and that 'expert tracking is essential to successful persistence hunting'.[2] Although tracking is often associated with men, it has been found that in the few modern hunter-gatherer societies in Southern Africa women also play roles in the process such as by cutting short foraging trips to report signs of game and then butchering dead animals after a hunt.[3]

Over the past few decades literature on tracking has taken two forms. First, there is a surprisingly large number of well-illustrated technical 'how to' books written by hunters, wildlife conservationists or former military Special Forces operators. Some of the military or law enforcement oriented ones contain very brief historical overviews of the use of tracking in warfare and promote continued training of soldiers in this field given the recent involvement of American and British forces in 'asymmetrical' warfare in Iraq and Afghanistan.[4] Secondly, a few books advocate the practice of tracking, which is stereotypically associated with North American indigenous people, as a way for modern humans to recover their lost relationship to the natural environment.[5] Both these genres of writing on tracking are connected to the rise in popularity of various types of wilderness tracking schools in the United States, Britain and Southern Africa – indeed, many of the authors operate their own schools – some of which are geared toward law enforcement, search-and-rescue and military applications, and others are meant for eco-tourists who would like a more intimate experience with wildlife. Recently, tracking and trackers have also featured in several feature films including Australia's 'The Tracker' in 2002, America's 'The Hunter' in 2003 and New Zealand's 'Tracker' in 2011, and several 'reality' television programs that involve wilderness survival and hunting including Canada's 'Mantracker' in which contestants try to elude an expert tracker.

Between the late 1940s and 1990, given the recent background of the Second World War and fears that the global Cold War between the United States and Soviet Union would turn hot, most Western national militaries focused on training and structuring their forces for large scale conventional warfare that was predicted to happen in Europe. With some exceptions, however, most wars fought in this period were characterized by guerrilla campaigns fought by emerging local nationalist movements bent on expelling European colonial rulers, other foreign occupiers or their proxies, and overthrowing minority settler regimes. Military forces which had been prepared to fight major conventional battles in Western Europe, reminiscent of the Second World War, were now given the job of hunting down small and elusive insurgent groups in faraway tropical environments. Usually following the popular revolutionary doctrine of Chinese communist leader Mao Zedung, insurgents usually did not wear uniforms and concealed themselves among the civilian population from whom they tried to gain support and recruits by persuasion or intimidation. Their operations usually began with small hit-and-run raids or ambushes on weakly defended state or economic infrastructure and they avoided encounters with large security force units that could defeat them in a pitched battle. Frequently, insurgents hid in remote forest, jungle or mountain areas. They hoped to slowly build up enough popular support and military resources to eventually challenge the weakened and demoralized state forces in conventional warfare and seize the centres of power, or to make the guerrilla conflict so costly and frustrating that their enemies would give up. Maoism was not the only revolutionary theory but, in practice, even insurgent groups that officially followed doctrines such as Leninism which advocated the sudden seizure of power centres by a professional corps of revolutionaries, fought hit-and-run guerrilla wars. As European powers withdrew from their colonial territories during the 1950s and 1960s, and former colonies became independent states, another feature of late twentieth century revolutionary warfare involved insurgents establishing staging areas in neighbouring countries with sympathetic governments where they could access material support and training often from Eastern Bloc powers. Rooted in the experience of fighting late nineteenth and early twentieth century colonial wars, counter-insurgency theories differed on the best response. Some advocated the use of overwhelming force to cordon off areas where guerrillas would be systematically hunted down and eliminated, and the imprisonment of large sections of the local civilian population in guarded camps to prevent them from assisting the insurgents. The French campaign against Algerian nationalists during the 1950s is usually held up as the classic example of this 'maximum force' approach. Others suggested that while it was still important to locate and engage the guerrillas with police-style 'minimal force', it was best to prioritize winning the 'hearts and minds' of the civilian community through economic and social development programs. The civilian population would then, as the theory maintains, refrain from supporting the insurgents who, lacking recruits and resources, would eventually surrender or be hunted down. This model is often seen as best represented by British operations against Communist guerrillas in Malaya during the 1950s. Another aspect of counter-insurgency theory, informed by the experience of the French in Algeria and Americans in Vietnam, stressed the importance of state security forces employing indigenous soldiers given their knowledge of local geography, language and culture.[6] In reality, most counter-insurgency campaigns of the late twentieth century utilized all these features to some extent and the well-known British 'minimum force' approach is now mostly seen as a myth meant to cover up atrocities in places like 1950s Kenya.

Regardless of the official theory adopted by counter-insurgency campaign planners, finding elusive guerrillas in overgrown terrain or among civilian communities was always challenging. Tracking was obviously vital to locating and engaging insurgents, particularly in remote wilderness areas, and arguably best conducted by troops with extensive local knowledge. Typical situations where tracking could prove useful to counter-insurgency forces involved reaction units following insurgent trails from scenes of ambushes, raids, mine-laying or border crossings and more proactive patrols searching for their spoor in areas of suspected guerrilla activity. In many circumstances state counter-insurgency forces also employed dogs tracking by scent to pursue guerrillas. Apart from actually finding enemy tracks to follow, the main problem of counter-insurgency trackers was how to catch up to their prey who could be minutes, hours or days ahead and perhaps making for the sanctuary of an international border or a populated area where their spoor would be lost. Depending on local geography and available resources, security force trackers enhanced their speed by trying to move fast on foot, riding horses, motorbikes or cross-country trucks and flying in helicopters all of which made tracking more difficult. On the other hand, insurgents quickly became aware that they were being tracked and learned to practice anti-tracking techniques which attempted to conceal signs of their movement and discourage pursuers by preparing an ambush or laying landmines on their trail. However, the conventional military forces involved in most Cold War era counter-insurgency operations usually struggled to find personnel with tracking skills and then to effectively utilize them within existing structures and tactics. According to British counter-insurgency practitioner and theorist Frank Kitson:

> Of all the specialist activities relevant to the prosecution of a counter-insurgency campaign none is more important than the provision of trackers ... Unfortunately, it is practically impossible to train a man to track in an acceptable length of time if he has no experience of it, and it is equally unfortunate that no indigenous trackers exist in many of the countries in which counter-insurgency operations may be expected to take place.[7]

Such realizations, which happened at the outbreak of insurgency in British administered Malaya following the Second World War, contributed to the establishment of the British Army Jungle Warfare School (JWS) in that country in the 1950s that standardized small unit training and included tracking and use of indigenous trackers such as the Iban of Borneo who came from a remote area and had a reputation as warrior head-hunters. The formation of Britain's JWS was also informed by the Second World War experience of fighting jungle campaigns against the Japanese such as in New Guinea and Burma where tracking had also been useful. Although Britain was not engaged in the war in South Vietnam, during the second half of the 1960s the JWS trained 1600 South Vietnamese and 200 American soldiers in subjects such as tracking. To supplement human visual trackers, the JWS trained American soldiers as dog handlers and the British Army supplied tracker dogs specifically for jungle conditions. Conventional American combat units in Vietnam were having trouble re-establishing contact with the enemy after an engagement when the insurgents appeared to melt into the jungle. Among the JWS instructors in the late 1960s were members of the New Zealand Special Air Service (SAS) who had experience of counter-insurgency warfare in Malaya. Also present was Major Huia Woods, a former New Zealand SAS operator now in the British Army, and 'one of the foremost jungle trackers in the world'.[8] An American soldier who attended JWS in the late 1960s prior to deployment in Vietnam later wrote 'The final exam for the course was a two week patrol in the dense jungles of Malaysia, tracking and being tracked by elite and fearsome Gurkha troops'.[9] It appears that after their traumatic experience in Vietnam, American forces abandoned serious counter-insurgency training during the last three decades of the twentieth century. However, recent combat against guerrillas in Iraq and Afghanistan has prompted the United States Army and Marine Corps to reintroduce combat tracker courses.[10] Despite the importance of trackers and many very brief mentions of them in the literature on counter-insurgency, very little has been written on the history of how tracking was employed during counter-insurgency campaigns.

In the late nineteenth century most of Africa was invaded and conquered by European powers motivated by the search for natural resources to fuel new industries, a sense of national pride and competition, beliefs in racial superiority and strategic considerations. Called the 'Scramble for Africa', this relatively rapid subjugation of a continent was facilitated by superior European technology in weaponry, transportation and communication, and the division and limited world view of African leaders. Completely new international borders were imposed on Africa and Africans, and during the early and middle twentieth century the continent was profoundly changed in political, economic and social terms by the experience of European colonial rule. Developments during the Second World War, including dramatic urbanization and rising expectations for national self-determination, led to the fast evolution of formerly moderate and narrow Western-style African political organizations into radical mass independence movements. These were encouraged by the successful independence campaigns in parts of Asia during the late 1940s and 1950s, the emergence of a bipolar international system dominated by two officially anti-colonial superpowers, and the growth of popular anti-colonial sentiment around the world including within the United Nations. In the late 1950s Britain, which had been engaged in some colonial reforms before the Second World War, decided to withdraw from its African territories as it was simply too expensive and politically embarrassing to hang on to them. The most important events in this process were the failed British and French intervention in Egypt in 1956, and the British campaign to crush the Mau Mau rebellion in Kenya from 1952 to 1956. Although France and Belgium initially tried to stave off calls for decolonization, they both eventually succumbed to international and local African pressures but withdrew from Africa in ways that attempted to continue their dominance after independence.

During the late 1950s and 1960s most African colonies were transformed into new independent states, initially modelled on Western democracies, through negotiation between outgoing European rulers and emerging local nationalist politicians. While there had been some incidents of violence during pro-independence protests, this political transition was mostly a peaceful process though the results would often result in terrible conflict in the future. Relatively peaceful decolonization was common in African colonies that had been administered by a relatively small number of European officials and where tropical climate had restricted European settlement such as those in West Africa. However, in the few territories where minority European settler societies had taken root and in some cases gained exclusive political power, such as Algeria, Kenya, Southern Rhodesia (today's Zimbabwe), South West Africa (today's Namibia) and South Africa, armed insurrections characterized the struggle for independence and/or majority rule. In these areas African people had serious grievances related to the imposition of a racially hierarchical state and economy, and the refusal of local settler leaders to contemplate reforms. Similarly, Portugal's fascist dictatorship refused to follow the example of other European powers by decolonizing African territories which meant nationalist guerrillas launched rebellions in Guinea-Bissau, Angola and Mozambique in the early 1960s. The strain of fighting three wars in Africa led to a military coup in Portugal in 1974 which caused the sudden granting of independence to these territories. Between 1952 and 1956 the British colonial administration in Kenya imposed a state of emergency in response to an African uprising called Mau Mau fought mainly by impoverished and landless members of the Kikuyu ethnic group that had been disproportionately impacted by white settlement. While the Mau Mau movement was eventually crushed by overwhelming force, the cost and embarrassment of the insurgency became an important factor in prompting a hasty British withdrawal from Africa including Kenya in 1963. The white settler dominated region of Southern Africa became the scene of many interrelated insurgencies and counter-insurgencies from around 1960 to 1990 when Namibia gained independence from apartheid South Africa which itself began a negotiated transition to non-racial democracy. The white minority regime in Rhodesia, after unilaterally declaring independence from Britain in 1965 over a dispute about proposed political reforms, fought a counter-insurgency campaign throughout the late 1960s and 1970s against several African nationalist revolutionary movements based in neighbouring black ruled countries and with Soviet and Chinese support. In South West Africa, from the late 1960s to 1989, insurgents fought for independence from the white dominated South African administration. The war in South West Africa spread into Angola as South African forces pursued insurgents to their staging areas in that former Portuguese territory itself involved in a civil war which was also a major proxy conflict of the global Cold War.

This book looks at the role of tracking in the decolonization era conflicts fought in the white settler dominated territories of Kenya, Rhodesia and South West Africa. It would be incorrect to assume that tracking is a static skill which was always employed the same way to simply hunt down insurgents. Over time, various methods, training programs and organizational structures were developed to enhance the practice of tracking in a military context. In Africa, this was informed by established colonial stereotypes about tracking and African identity, and the previous employment of trackers during wars of colonial conquest. The use of this capability by state counter-insurgency forces clearly and purposefully evolved from one conflict to the next and was influenced by factors such as local environments, security force culture, technology, resources and insurgent tactics. Key individuals, often with extensive hunting or game-keeper experience, were important in developing these initiatives. On the other side, the insurgent use of tracking and anti-tracking techniques appears to have been mostly reactive but it was not always the same and was influenced by numerous factors such as insurgent background and the natural environment. Tracking represents an important yet only vaguely understood element of security force and insurgent operations as well as the daily experiences of combatants during these conflicts. However, it was always part of a wider set of circumstances and policies, and should not be seen in isolation. In a sense, this book looks at the history of how one particularly important tool in the counter-insurgent – and to some extent insurgent – toolbox was used in a specific period and region. One of the main reasons for writing this book is to provide students of today's 'asymmetrical wars', which tragically seem set to continue well into the future, with an in depth look at the history of how tracking was utilized in previous similar situations. The focus on decolonization conflicts in African territories with white settler minorities is related to my own expertise. Furthermore, there is value in looking at the specific theme of tracking within the broader framework of a shared regional history involving issues such as settler colonialism, African nationalism and a broadly common British security force culture. There are several other regional factors that require some explanation. Given that military service was seen as linked to the possession of citizenship rights and dominant white minorities feared arming blacks, the local defence forces of colonial Kenya and Rhodesia, and apartheid South Africa relied on the conscription of young white men. Black men could volunteer for security force service but they worked at the bottom of a clear racial hierarchy. In addition, during the early colonial era black communities had been evicted from large parts of these territories which were demarcated as game reserves for exclusive use by white hunters and in which hunting by blacks was criminalized as poaching. As a new philosophy of complete nature conservation arrived from North America, some of these reserves were declared national parks in the late 1940s and 1950s at the same time as radical African nationalism was emerging. Conservation laws were enforced by wildlife and eventually national parks departments the organization of which, like the police and military, reflected the racial hierarchy of colonial society as whites occupied senior positions as armed game rangers or wardens and blacks were unarmed game scouts. During Africa's decolonization wars, elusive guerrillas often sought sanctuary in isolated game reserves or national parks, and personnel from these institutions were frequently mobilized by state security forces. This led to the militarization of African national parks the legacy of which endures to this day.[11] As I lack proficiency in Portuguese, the role of tracking in Portugal's decolonization wars will not be examined in detail though it will be discussed when related to events in Rhodesia and South West Africa both of which bordered Portuguese territories.[12]

There does not appear to be a wealth of documentary evidence related to the military utilization of tracking in these conflicts. State forces usually began to adopt tracking on an ad hoc basis and many trackers were ill-literate people from marginalized communities. The British army in Kenya did maintain some files of correspondence and records related to the development of a tracking school and tracker combat teams which are accessible in the National Archives, London.

Some of these were opened to the public as recently as 2007. Unfortunately, the Rhodesian Army Association Archives, a collection of security force documents spirited out of Zimbabwe on the country's independence in 1980 and briefly opened to researchers at the now defunct British Empire and Commonwealth Museum in Bristol during the late 2000s, has been closed over concerns about confidentiality and access. Similarly, much of the official South African documentation related to military operations in South West Africa in the late 1970s and 1980s remains classified. As such, the many and often recently published personal memoirs of security force veterans from the conflicts in Kenya, Rhodesia and South West Africa, aspects of which several historians have described as a 'minor publishing industry', constitute particularly important sources for this book.[13] While almost all of these accounts were written by white veterans who observed black trackers, interviews with some surviving black security force veterans of the 'Bush War' in Rhodesia and the 'Border War' in South West Africa has been useful. There is only one published account by an African security force tracker who fought in one of these conflicts and he has now passed away.[14] Information on tracking or anti-tracking from the insurgent perspective is more difficult to find particularly as these guerrilla organizations tended to avoid keeping records for both practical and security reasons, the former insurgent movements now in power in Zimbabwe and Namibia jealously guard their closed archives, and many insurgent veterans lack the resources or motivation to write their stories. However, a small and growing number of published accounts by former insurgents from Kenya, Zimbabwe and Namibia have been invaluable in writing this book. Of course, accounts by veterans of the different sides of these wars must be read critically as most seek to vilify their former enemies or justify wartime actions.

1 TRACKING AND IDENTITY

The involvement of African trackers in colonial warfare and later counter-insurgency campaigns was informed by the way that colonial Europeans associated tracking with specific African identities. In colonial stereotypes different African ethnic groups were often linked with supposedly innate traits, skills or jobs such as Zulu, Ndebele and Maasai who were seen as noble warriors, Mfengu who were seen as industrious and educated, Mpondo or Lamba who were believed to be wild, Nyamwezi who became caravan porters and Bhaca who were hired as night-soil carriers in South Africa's mines. Within a divide-and-rule context, some of these stereotypes presented neighbouring peoples as polar opposites such as Southern Rhodesia's honest and brave Ndebele versus the duplicitous and cowardly Shona or South Africa's martial and upright Zulu versus the treacherous and criminal Xhosa. While African ethnic stereotypes are popularly seen as timeless, scholars often see them as having emerged in the competitive economic atmosphere of colonial towns and mining centres where they were encouraged by opportunistic European and African elites as a way to mobilize specific groups.[1]

One of the most frequently repeated and enduring ethnic stereotypes related to Africans is that which ascribes superior tracking and hunting skills to marginalized minorities such as the Bushmen of the Kalahari region, the Ndorobo of Kenya, the Shangaan of the Zimbabwe/Mozambique/South Africa border and the Twa of Central Africa. Such beliefs emerged simultaneous to the rise of European sport hunting in Africa during the late nineteenth century. While it has been pointed out that tracking was one of the few areas in which colonial Europeans acknowledged superior African skill,[2] those African groups perceived as extraordinary trackers were also the most disdained in almost every other respect. In the colonial mind, skill in tracking came to be very strongly associated with African minority groups who tended to live by hunting and gathering in remote wilderness areas. The many African groups that historically lived in large settled communities, formed organized states, practised mining and metallurgy, cultivated crops and raised livestock were seen as alienated from the natural environment. Hunter-gatherer minorities were perceived to have a much closer relationship with the environment or indeed as part of nature itself. By the time of colonization, relations between some the larger agricultural groups and neighbouring small hunter-gatherer communities had often become unequal and exploitative. The perceived primitiveness of these hunting communities and sometimes the small physical stature of their members meant that from a colonial Social Darwinist perspective they were looked down upon as occupying the lowest position in the hierarchy of human races or perhaps not completely human. Since they were perceived as similar to animals, these marginalized minorities were believed to possess superior beast-like senses and abilities associated with hunting that could be tamed and utilized by more civilized people. It was believed that the more primitive a community was then the better its members would be at tracking. Of course, not all African groups associated with tracking excellence lived as pure hunter-gatherers though they were always marginalized in some way particularly in terms living far from major centres or their association with other supposedly primitive ways of life such as pure pastoralism. The advent of colonialism furthered the alienation of hunter-gatherer minorities. As the majority African communities became incorporated into the colonial capitalist economy as cheap labour or peasant farmers, and missionary western education spread, this perceived gulf between the supposedly primitive trackers and all others widened. Colonial observers were often so impressed by the talent of some of these marginalized peoples to follow game tracks that many such groups were declared the best trackers in the world. Colonizers also saw these minorities as potential allies in the struggle to dominate the larger groups and this special relationship, at least within the colonial imagination, developed during the growth of European sport hunting in Africa and was popularized by hunting literature. According to anthropologist David McDermott Hughes, 'Hunting narratives almost invariably associate loyal African trackers – from minority non-agricultural tribes – with equally devoted Great White Hunters.'[3] As we will see, this imagined relationship would continue into the decolonization wars of the late twentieth century. John Taylor, a controversial Irishman who hunted in East Africa in the early to middle twentieth century and who dedicated one of his books to his Malawian tracker and gun bearer Aly Ndemanga, placed African tracking prowess within the context of hunter-gatherer communities elsewhere in the world:

> The Australian Blackfellow has the name of being the best tracker in the world, and police out there would be pretty well lost without him. The Sind trackers retained by the police in Peshawar [Pakistan] also have a very fine reputation. But I honestly don't think that either are better than really good Africans. Possibly the Australians, taken as a whole, are better than the general run of Africans; but then the Australian is still back in the stone age and relies solely upon his hunting, whereas the majority of Africans are agricultural or pastoral, not, generally speaking, hunters. The only tribes known to me who live on the proceeds of their hunting are the Kalahari Bushmen, the Pygmies of the Ituri Forest, and the Wanderobo in Kenya and Tanganyika. Accordingly, it's unfair to take the African generally and compare him with the Australian aborigines.[4]

The most popular contender for the title of world's best trackers was and remains the Bushmen (also called Khoisan, San or Basarwa) of Southern Africa's Kalahari Desert region which includes part of present day South Africa, Botswana and Namibia. This reputation is related to the Bushmen's historic hunter-gatherer way of life which involved considerable tracking and hunting in a harsh environment. Such ideas were also associated with the linguistically similar Khoikhoi pastoralists, called Hottentots in colonial times, of the Cape. Historian John MacKenzie explained how nineteenth century Europeans were captivated by Khoikhoi and Bushmen hunting techniques. 'Their incomparable eye for tracking and capacity to stalk their prey when it was sighted led to their being employed as ideal auxiliaries for European hunters'.[5] Recent environmental historians of the British Empire contend that 'Khoisan tracking skills were legion, and their ability to read the signs of the earth construed almost as a form of literacy'.[6] While this is certainly an accurate reputation of the dominant colonial view, the stereotype of the skilled Bushman tracker did not develop immediately upon their first colonial encounter and has been the subject of some dissenting views.

From 1843 to 1848 Roualeyn Gordon-Cumming, a British soldier who pioneered sport hunting in Southern Africa, traveled throughout the area of present-day Botswana and acquired local guides from missionary David Livingstone. It is likely that before embarking on his professional hunting career, Gordon-Cumming had acquired some tracking knowledge while hunting in India and as a junior officer in the Cape Mounted Rifles (CMR) which patrolled the Eastern Cape frontier. Eventually celebrated as 'The Lion Hunter' who's writing popularized European sport hunting in Africa, Gordon-Cumming did not generally associate tracking with Bushmen or any other ethnic group and saw it as skill held by specific and rare individuals. Throughout most of his five years of hunting in Southern Africa, Gordon-Cumming employed a young Bushman named Ruyter who, as a child, had been captured by Boers among whom he learned to ride horses which would become a useful skill for a hunter's assistant.[7] Ruyter did many jobs including tracking with Gordon-Cumming which was always presented as a cooperative activity, driving game out of dense bush, and serving as a messenger and guard. Although he tracked and hunted many animals with Ruyter, Gordon-Cumming did not romanticize his tracking skills or associate them with his ethnic origin. Indeed, the trait that Gordon-Cumming most valued in Ruyter was his loyalty which prompted 'The Lion Hunter' to bring him back to Britain.[8] When entering a new area, Gordon-Cumming always employed local trackers for a short period and most of these were Tswana men from large settled chiefdoms which in later colonial stereotypes were not considered good trackers. He described one of his early tracking experiences:

> I saddled up, and proceeded to take up the spoor of the largest bull elephant, accompanied by after-riders and three of the guides to assist in spooring. I was also

accompanied by my dogs. Having selected the spoor of a mighty bull, the Bechuanas [Tswana] went ahead, and I followed them. It was extremely interesting and exciting work. The foot-print of this elephant was about two feet in diameter, and was beautifully visible in the soft sand. The spoor at first led us about three miles in an easterly direction, along one of the sandy footpaths, without a check. We then entered a very thick forest ... After following the spoor some distance farther through the dense mazes of the forest, we got into a ground so thickly trodden by elephants that we were baffled in our endeavours to trace the spoor any farther; and after wasting several hours in attempting by casts to take up the proper spoor, we gave it up, and with a sorrowful heart I turned my horses head toward camp.[9]

Remembering finding the tracks of two large elephants at a waterhole, Gordon-Cumming wrote 'This was glorious! I had great faith in the spooring powers of the Bamangwato men, and I felt certain that at length the day had arrived on which I was to kill my first bull elephant. The Bechuanas at once took up the spoor, and went ahead in a masterly manner; and with buoyant spirits I followed in their steps'. The Bamangwato were from one of the larger and most northerly Tswana states, the Ngwato, which would later in the century court good relations with the British to stave off aggression by their Boer and Ndebele neighbours. Gordon-Cumming's hunting party eventually lost the elephants' trail when they passed through an area of bush where many other elephants had eaten berries and therefore disturbed the ground.

After a fruitless search of several hours, and many vain endeavours to retrieve the day by trying back on the spoor and making wide casts to the right and left, I was completely beaten and compelled to drop it, the Bechuanas sitting down and sulkily refusing to proceed farther.[10]

'The Lion Hunter' described a later and more successful elephant hunt with 'the native who led the spooring party being the best tracker in Bamangwato'.[11] On another occasion Gordon-Cumming was aided by the Tswana chief and expert hunter Seleka and 150 of his men in tracking elephants whose presence in a forested area had been reported by some Bakgalagadi hunter-gatherers who, although not Bushmen, have also acquired a reputation for tracking. The Irish hunter wrote that

The spooring was conducted very properly, the old chief taking the greatest care of the wind, keeping his followers far back, and maintaining silence, extending picquets in advance, and to the right and left, and ordering them to ascend to the summits of the tallest trees to obtain a correct view of the surrounding forest. Presently, the mighty game was detected.[12]

In the late 1850s, as a hunting economy expanded in the Kalahari region with products such as ivory and hides being exported to the coast, British hunter William Charles Baldwin ventured into many of the same areas as Gordon-Cumming and employed one of the same African guides called Kleinboy. Unlike his predecessor, Baldwin much admired the tracking skills of Bushmen and depended on them to locate water sources. He was also one of the first colonial hunter-writers to begin comparing them to bloodhounds which would be copied by many subsequent authors in the genre. Just as the superior olfactory senses of a bloodhound could be put to use by his human owner to facilitate a hunt so too could the more developed visual tracking skills of the Bushmen. During a search for a horse that had wandered from Baldwin's camp:

> Inyous and one Bushman Kaffir did most of the hunting. Once, I had all but given him up, on flinty, rocky ground; we cast around in every direction for an hour and a-half to no purpose, and we followed the spoor for more than 300 yards on our hands and knees, the faintest imaginable track being all we had to guide us – a small stone displaced or a blade of grass cut off; so we kept on till we again got to sandy ground, when we took up the running about four miles an hour; and about mid-day we found him ... I must say that today's work beat anything I ever saw with Kaffirs. Bloodhounds could not have done better. We followed the trail for six hours through old grass a yard high, and through the midst of lots of quagga spoor. I once called the Kaffirs to a quagga spoor, but they recognized it immediately, and made me ashamed of myself.[13]

Coercion was sometimes used to secure the services of Bushman trackers. On one occasion Baldwin's hunting party surrounded a Bushman settlement to prevent their flight and then shot some antelopes for them to entice them to work as guides. The hunting party then

> took up the spoor of the day previous ... the Bushmen, eight in number, followed it beautifully until about 1.a.m., when they got on spoor only a few hours old; the scent freshened wonderfully, and the Bushmen hunted to perfection, their captain taking the lead throughout, and being infinitely the best man I ever saw.[14]

Describing a successful elephant hunt in a forest, he stated that his tracker

> January hit off the fresh spoor of an old bull, followed it ten times better than the best blood hound over all kinds of country, hard and soft, lots of open plains, and dense 'vac um bechis' [thorn bushes], at a rattling pace, without an instant's check, hour after hour.[15]

Baldwin praised 'the wonderful instinct of January, who is the finest hand at a spoor I have ever heard or even read of'.[16]

From the late nineteenth century, the seemingly amazing tracking skills of the Bushmen became a standard feature of the many books written by European and North American sport hunters who visited Southern Africa. This was perhaps influenced by the emergence of pseudo-scientific racial theories such as Social Darwinism which presented humanity as divided into a hierarchy of races. The Bushmen were clearly seen as occupying the bottom rung of the racial ladder and, therefore, their perceived superior skills in tracking and bush navigation were portrayed as part of their semi-animal nature. This was also the period of Southern Africa's mineral revolution during which many large African communities became embroiled in migrant labour to the mines which, in the colonial mind, further divorced them from nature which made isolated hunting minorities seem even more exotic and rare. During the 1870s Czech explorer Emil Holub wrote that 'No people in South Africa are more skilful than the Masar was in foraging out water in dry districts, or more keen in tracking game'.[17] In the same decade former British officer and professional sport hunter Parker Gillmore was greatly impressed by the Bushmen. 'These Kalahari bush-people are the most persevering and courageous hunters; once on the trail of game, they never leave it till they kill; and their skill in stalking cannot be surpassed'.[18] Describing how three Bushmen showed him how to engage in the persistence hunting of some zebra and giraffe, Gillmore wrote that

> For a quarter of an hour the Massaras (sic) displayed their skill and perseverance as trackers. Not at a walk did they follow the game, but at a run, so swift that they kept Ruby [Gillmore's horse] at a good round trot, and so unerring that not for a moment did they appear at fault ... Wonderful fellows these bushmen – they rival the snake in its subtlety, the eagle in its power of sight.[19]

Hunting in what is now Botswana in 1890, British sportsman H. Anderson Bryden employed Bushmen assistants who he believed were

> the most wonderful trackers, perhaps in the world. It is a fact that a Masarwa can, from the appearance of the spoor, tell you to within a few minutes how long it is since game has passed. Their instinct in this respect, and the faculty of finding their way in the wildest veldt is quite unerring.[20]

Frederick Courtney Selous, the late nineteenth century arch-type of the great white hunter who led the British occupation of Southern Rhodesia (now Zimbabwe) in 1890, specifically linked Bushmen wilderness skills with their supposed animal nature as he believed that their 'extraordinary faculty in finding direction' without landmarks was similar to 'oxen, horses and elephants'. He also thought that, like hunting dogs, it was best to keep Bushmen trackers hungry:

> But to return to the Masarwas. As trackers and assistants in the hunting veld they are unrivaled, and they are more docile and less assertive than Kafirs. To be seen at their best they must be hungry, and lightly indeed must the wounded animal tread that hopes to escape from a half-starved Masarwa, but as soon as they get fat, they become lazy and careless, like dogs. Their life is a hard one. Such seven-eighths-starved Bushmen are splendid fellows as assistants in tracking game.[21]

Selous's tendency to compare the Bushmen to animals may not have entirely originated in late nineteenth century European racial theories but also in local African opinions as he claimed the Ndebele King Lobengula had told him 'The Bushmen are not human beings; they are only wild animals'.[22] A more recent biographer of Selous continued these views by describing the hunter's half-Griqua, half-Bushman employee called Laer as having 'an animal-like instinct for tracking'.[23] Bryden and Selous went on to become founding members of the Shikar Club (Shikari is an Indian word for hunting) which promoted elite sport hunting as a way to reinforce masculine virility in the British Empire and eventually transformed into an early conservationist organization. It is very likely, given the above statements, that this club popularized the idea that Bushmen were ideal though sub-human trackers and guides essential for a successful hunting trip to Southern Africa.[24] Comparing Bushmen or Khoikhoi trackers to animals became common. Famous American scout Frederick Russell Burnham, a colleague of Selous during the conquest of Southern Rhodesia in the 1890s and chief scout for the British army during the Second Anglo-Boer War (1899–1902), wrote that

> It is almost as hard to shake a "Totty" [Hottentot or Khoikhoi] off a trail as a bloodhound. These yellow, beady eyed natives come from the southern deserts and are only a shade above the now extinct Bushmen in the human scale ... They made good trackers, herders and grooms and, like a good dog, they obey only one master.[25]

Arnold Wienholt, an Australian soldier, cattleman and politician, favourably compared the hunting and wilderness skills of the Bushmen he employed as trackers in Angola with the indigenous people of his homeland. He also compared tracking to literacy. According to Wienholt, 'Beautiful trackers are these Bushmen, the sand with all its footprints being their newspaper, which they read as they travel along'.[26]

European accounts of Bushmen tracking contain contradictions such as over how silent they would remain during a hunt. In the early twentieth century, Christian missionary and amateur ethnographer Samuel Shaw Dornan wrote that the Bushmen of Bechuanaland (today's Botswana) were courageous hunters who would track a lion or leopard and take the prey it had been stalking. Furthermore, he believed that

> As trackers they are simply invaluable. Nothing escapes their notice: a piece of bent grass, an overturned pebble, or a broke twig is quite sufficient to tell them where the game is gone and how far it is off. They will keep on the spoor of a wounded buck for hours, usually without speaking a word, and then, when the animal is sighted, they will quietly point it out to the hunter ... They are as wary as the animals they stalk.[27]

A British official posted to south-western Northern Rhodesia (Zambia) on the border of German South West Africa's Caprivi Strip during the First World War found his district inhabited by a small community of Makwango Bushmen who 'were by far the best trackers I had ever hunted with'. Impressed by the Bushmen trackers who walked along laughing among themselves and rarely looked at the ground when hunting a dangerous lion, he wrote 'I have never seen spooring to equal it'.[28]

From the mid twentieth century, academics and film makers took over from
sport hunters as the main celebrators of Bushman tracking finesse. Archaeologist
John Desmond Clark, leader of an expedition from the Livingstone Museum
that visited the few remaining Hukwe Bushmen in western Northern Rhodesia in the late 1940s, observed that 'The bushmen are the finest trackers in the
world and will follow for hours and days an animal wounded with one of their
poisoned arrows'.[29] During the 1950s American ethnographer Lorna Marshall
observed that Bushmen

> Hunters memorize visual impressions and are able to follow the tracks of an indi-
> vidual animal in the midst of a large herd in a way that seems to us miraculous ... They
> register every person's footprints in their minds, more vividly I am sure than we do
> faces, and read in the sand who walked where and how long ago.[30]

Elizabeth Marshall Thomas, the daughter of Lorna Marshall who lived with the
Kalahari Bushmen in the 1950s and did much to mythologize them as innately
peaceful people, wrote that

> The tracking ability of the Bushmen is legendary, and rightly so. I happened to be
> traveling with three Ju/wa men who had occasion to track a hyena across a wide slab
> of bare rock. How they did it I have no idea. They were not simply following the line
> of travel, because out on the rock, the route of the hyena made a curve of about one
> hundred degrees and emerged about one hundred feet away on sandy ground on the
> far left of the ledge at a place that, I, for one, had no reason to anticipate. The three
> men had anticipated it, however, perhaps because of the heavy bushes that grew there
> ... The feat seemed effortless.[31]

For anthropologists Phillip Tobias and Megan Biesele, Bushmen had evolved
diminutive physical size because they did not need great strength to handle their
small bows as 'the San depends rather on his fleet-footedness and extraordinary
tracking skills'.[32] Richard Lee, another anthropologist, saw Bushman tracking
skill in more practical terms:

> The !Kung are such superb trackers and make such accurate deductions from the
> faintest marks in the sand that at first their skill seems uncanny. For example, both
> men and women are able to identity an individual person merely by the sight of his
> or her footprint in the sand. There is nothing mysterious about this. Their tracking is
> a skill, cultivated over a lifetime, that builds on literally tens of thousands of observa-
> tions. The !Kung hunter can deduce the following kinds of information about the
> animal he is tracking: its species and sex. Its age, how fast it is traveling, whether it is
> alone or with other animals, its physical condition (healthy or ill), whether and on
> what it is feeding, and the time of day the animal passed this way.[33]

Laurens Van Der Post, through his writing and particularly the 1950s television
documentary series 'Lost World of Kalahari', did much to popularize the concept
of Bushmen as innate trackers and hunters. The 1958 documentary film 'The
Hunters', made by John Marshall, son of Lorna and brother of Elizabeth, focuses

on a small group of Ju/Wasi Bushmen in northern South West Africa who shoot a giraffe with a poisoned arrow and then experience the exhausting process of tracking the fleeing animal across dry and hard ground with few obvious signs and eventually kill it with spears. Unlike Van Der Post's television series, Marshall's film did not portray Bushmen as possessing inborn tracking abilities and described individuals with different levels of hunting skill. Overall, these films popularized a visual image of the Bushmen as pristine hunter-gatherers living in a desert Eden untouched by western civilization. Tracking represented a central feature of this image.[34] A popular view developed that all Bushmen were born with advanced tracking and orientation skills. They were portrayed as 'superhuman, or rather "superbeastly"'.[35] Some South African scientists explained superior Bushmen tracking skills in terms of better visual acuity.[36] Today's ecotourism industry in Botswana builds upon this reputation by using the internet to advertise safaris that include 'spending time with the Bushmen, learning about tracking and the secrets of Botswana's Kalahari Desert'.[37]

Not everyone agreed with the hunting and tracking aspect of the Bushman myth. One of these was G. B. Silberbaurer, a colonial official in Bechuanaland who conducted a survey of the Bushman population in the 1950s that eventually led to the creation of the Central Kalahari Game Reserve (CKGR) to preserve their hunter-gatherer way of life though it confined them to a marginal area. He maintained that

> the Bushmen are not the great hunters legend has them to be. They are, in the first place, thrifty in their hunting and do not shoot more than enough for their requirements. Although skillful trackers, they are not infallible stalkers or marksmen and often go for weeks without meat.[38]

Furthermore, he observed that the G/wi Bushmen of the Kgaotwe Pan area 'are really rather poor hunters and only do well when game is present in large numbers and not easily disturbed'.[39] Peter Hathaway Capstick, an American professional hunter in Rhodesia and Botswana in the 1970s who wrote a series of popular hunting books, did not think that all Bushmen were born trackers:

> Usually, in a family group, there will be one or two men who are responsible for hunting, and they are among the finest trackers imaginable. Civilized man has credited all Bushmen with great hunting skills, but, in truth, most are merely gatherers and scavengers, unable to track a wounded hippo through a fresh snowbank. I have usually had problems hiring pure Bushmen as trackers because, living so close to the land, they would only work long enough to earn some beloved tobacco and, after a few days, be gone without warning.[40]

However, Capstick still romanticized Bushman trackers such as when he wrote that 'When Debalo tracked leopard, he didn't think like one, he *was* a leopard'.[41] For travel writer and Botswana resident Mike Main, the skill of a particular Bushman tracker he observed was explained by a variety of factors:

Bom's knowledge of his environment is extraordinary. It encompasses not only the behaviour and characteristics of the fauna and flora in his environment but also shows a profound understanding of the manner in which they interact with each other and with the environment itself. He is drawing strands of expert knowledge from a number of different fields and, from this, drawing his remarkable conclusions.[42]

In East Africa, Ndorobo hunter-gatherers were labelled as outstanding trackers. They were often employed in this capacity by European hunters and farmers of the colonial era, and the Ndorobo tracker became a stereotypical character in many hunting stories set in Kenya. For early twentieth century hunter William Robert Foran, Kenya's Ndorobo

without any doubt, are far and away the most proficient trackers in all Africa. It is probable that an Ndorobo hunter is without a peer in this difficult art. I have tried many men of various tribes, but have never found one that can touch the skill of a Ndorobo.[43]

Emphasizing their primitiveness, Foran called the Ndorobo 'caveman trackers'.[44] Another admirer of Ndorobo tracking abilities was former United States president Theodore Roosevelt who famously led a Smithsonian Institute expedition to East Africa in 1909 and 1910 to hunt animals for museum exhibits and was accompanied by Selous who he idolized. Of the Ndorobo, Roosevelt wrote 'Their eyesight was marvellous, and they were extremely skilful alike in tracking and seeing game. They threaded their way through the forest noiselessly and at speed, and were extraordinary climbers'.[45] Elspeth Huxley, the famous white Kenyan author, wrote that her servant 'Njombo was far more expert, but even he was not a tracker as the Dorobo people were and many of the Wakamba'.[46] A late 1960s East African travel guide reported that

The Ndorobo live in the deep forests or the open bushlands, shunning contacts with the outside world, living with the animals so that they have an ability to track and understand them that appears magical to the European and even other African tribes.[47]

The biographer of early twentieth century hunter Denys Finch Hatton wrote that 'Like other diminutive hunters, the Dorobo, who haunted the forest fringes, were despised by taller tribes. They were the best trackers in East Africa and could skin a buffalo in five minutes'.[48] Others debated the issue. William Stephen Rainsford, an American who hunted big game in Kenya in the first decade of the twentieth century, wrote that 'Some experienced hunters advise the engaging of Ndorobo trackers and say there are none so good. I have found the Wakamba to be about the best trackers in the country'.[49] Captain C.H. Stigand, who believed all Africans were inferior hunters and trackers, was more sceptical of Ndorobo tracking skills as

The Nandi and Wandorobo, or Ogieg, hunters in the forest do a certain amount of tracking, but in the bush or anywhere outside of their own forests they appear to be quite useless. In the forest no great skill in required, as the tracks are all deep and plainly obvious.[50]

Early Kenya game warden A. Blayney Percival was also pragmatic about African tracking skills and wrote 'Primitive races are popularly supposed to be particularly gifted in this regard ... indeed, among the East African races the skilled tracker is a distinct rarity'.[51]

In the tri-border area of south-eastern Zimbabwe, eastern Mozambique and north-eastern South Africa, the Shangaan people have acquired a reputation as accomplished trackers. Although they were not historical hunter-gatherers but rather settled farmers and herders, the Shangaan live in a distant frontier zone. James Stevenson-Hamilton, the first chief warden of South Africa's massive Kruger National Park, associated different African ethnic groups with specific skills such as the Swazi who he believed were excellent cattle herders, the Sotho who he thought liked to work as farmer labour or wagon drivers, and the Tsonga (or Shangaan) who he saw as having great aptitude for mechanical work in mines and industry. In terms of hunting and tracking, he thought that environment and experience were most important:

> The Hlangane clan of the Thongas (Shangaan), having through many years been obliged to subsist mainly on the wild game which they could succeed in killing, still contains some good hunters. Many of the older men yet living are, in their knowledge of the habit of wild animals, in tracking lore, and in their skill in setting snares, comparable with any natives in the African continent. Individuals of the Ama-Mbayi and of the Lowveld ba-Swazi are also excellent hunters and trackers.[52]

Growing up in south-eastern Rhodesia during the 1950s and 1960s, award-winning conservationist Clive Stockil 'spent more and more time with particular individuals amongst the Shangaan community. With the best trackers, the most successful hunters. They were men I looked up to'.[53] In the late 1980s young game ranger Graham Cooke was new to Londolozi Game Reserve near Kruger National Park:

> I was assigned to work with a Shangaan tracker named Carlson Mathebula who shared with me his vast knowledge of all aspects of bush craft, from tracking wild animals to evaluating their spoor ... Carlson and I worked closely together for a number of years and what I found most impressive of all was his uncanny ability to track lions. He seemed fearless, picking up on and investigating the tiniest of clues such as finding a single lion's hair left behind on a small bush. He walked ahead of me but I still felt a little anxious being on foot close to a lion, even though I was the one armed with a rifle.[54]

While researching lion attacks on Mozambican migrants travelling through South Africa's Kruger National Park, journalist Robert Frump was told that 'Shangaan were "bush smart", said by some to be the world's best trackers'.[55] Working in the Kruger National Park as trackers and guards became a central feature of loyalist Shangaan identity which paternalistic park authorities contrasted with other local African groups such as the Makuleke who had been evicted from the reserve and were considered enemies of nature conservation.[56] Today, game parks and tourist tracking courses advertise that clients will be led by authentic Shangaan

trackers. The website of Simbavati River Lodge claims that clients will 'be aston-
ished by the legendary bush-craft and tracking skills of your Shangaan tracker'![57]
An American periodontist hunting big game in south-eastern Zimbabwe wrote
that 'Under difficult tracking conditions, Peter's skill and intuition kept us on
the trail of our quarry – the mighty elephant ... Shangaans are reputed to be the
best trackers in the world, and Peter upheld this'.[58] Academics are not immune to
such views. A recent University of Connecticut doctoral candidate in environmental studies explained that 'I wanted to go to Africa and learn tracking from
the Shangaan because they are the acknowledged masters of the art'.[59]

The Twa hunter-gatherers of the Central African forests, called pygmies in
colonial times, have been characterized as particularly gifted trackers but also
sub-human. During the early twentieth century, some Twa began to work as
trackers for European hunters and naturalists who wanted to locate elusive forest
animals such as elephants, bongo and mountain gorillas. In 1906 British hunter
P. H. G. Powell-Cotton spent 10 months in the Ituri region of the north-eastern
Congo Free State (today's Democratic Republic of Congo) where he was greatly
impressed by his Twa trackers. 'In all my wanderings, I have never met a native
race so adept at tracking or so thoroughly acquainted with every habit and haunt
of the animals of their country'.[60] Around the same time P.J. Pretorius, a Boer
elephant hunter originally from the Transvaal who spent three months with the
Twa in eastern Belgian Congo, 'found the pygmies splendid trackers and absolutely fearless'.[61] Similarly, naturalist and explorer Cuthbert Christy visited the
same area in 1912 and wrote a detailed account of Twa tracking:

> The intimate knowledge possessed by the pygmies of the private life of every animal
> great and small was a source of deepest interest to me. Their powers of tracking appear
> simply marvelous till one learns, by frequent association with them, something of the
> art oneself. At first I found it difficult to pick up the trail as it were, but once started
> they become wholly absorbed in the work and very pretty it is to watch them. They
> give their opinion as to age of tracks, fresh or otherwise, in a low whisper, and fol-
> low on with almost cat-like stealth, careful to make no sudden movement, peering
> in front but missing nothing on the ground. They are alert for the least sound ahead
> and communicate by almost imperceptible gestures, without looking round, all that
> they find to those behind them such as; here it was lying down, knees here, back there,
> here it has dropped those leaves. When two are working together the pace is quicker
> than with one. If the leader is at fault the smallest gesture tells the man behind in
> which direction to look. The front man from the corner of his eye tells instantly if
> the second man finds the track and he follows in that direction, and so on, first one
> leading, and then the other. If both are at fault one man trots ahead along the most
> likely spaces in the undergrowth, while the second man makes a sort of cast round
> where the tracks ceased, working like a ferret with nose to ground, turning over a leaf
> here or feeling with the flat hand there for any depression made by the hoofs, all the
> while on the alert and watching for any sign from the man in front if he can see him.
> If nothing is found on he goes, pretty sure that the other one has picked up the tracks
> ... Without the Mambutte the sportsman in these forests can do absolutely nothing.[62]

In the 1920s American hunter Ben Burbridge secured the services of Twa trackers to help him capture gorillas in the Virunga Mountains on the Belgian Congo/Rwanda border. While recruiting local porters and guides at a White Fathers' mission which had become the centre for such activity, Burbridge was sceptical about those who claimed to know how to find gorillas. However,

> Then came some little Twa men of a semi-dwarfish tribe who lived in the surrounding forest. These men are famous trackers, hounds of the forest who can follow unerringly any spoor and creep upon game with the stealth of a roach in rubber slippers.[63]

Italian explorer and writer Attilio Gatti believed that contact with Europeans had spoilt the simple and authentic Mbuti Twa of the Tchibinda forest of eastern Belgian Congo who, in the 1930s, he described as 'lazy, avid, ridiculous caricatures of men whose highest ambition is to ape the white in everything'. However, Gatti was pleased to find that Mbuti leader Kasciula, who he had met on a previous expedition, 'revealed himself as the authoritative chief, the magnificent tracker, the courageous hunter of four years before, each time during the following days that we found ourselves near the gorillas'. Furthermore, Gatti recognized his and other European visitors' dependence on 'the guidance of the extraordinary knowledge that the pygmies seem to have of even the intimate thoughts of a gorilla'.[64] In the 1950s an Oxford anthropologist wrote that

> While the other natives of the Ituri rely upon their ingenuity to trap elephant, the pigmies rely upon their skill as trackers and hunters. In the art of tracking they are expert and in the face of danger extremely courageous.[65]

During the late twentieth century various marginalized Twa groups were evicted from national parks and displaced from forests by commercial logging companies. In recent years, Baka Twa have been employed in parts of Cameroon, Gabon, Congo-Brazzaville and Central African Republic (CAR) as trackers for European professional hunters who take clients on forest hunts and eco-tourism operations.[66] According to Geoffroy de Gentile, a French professional hunter in Cameroon, 'The pygmies are the best hunters in Africa. They live in the forest. They know everything about the forest. They find tracks where there are no tracks'.[67] A gorilla safari operation in CAR advertised that

> This task could not be achieved without the help of pygmy trackers whose senses are perfectly in tune with the forest. The tracking experience alone, searching for the gorillas by following their almost imperceptible signs on the forest floor, leaves you with an unforgettable memory'.[68]

An Italian expert on western lowland gorillas in CAR wrote that

> Anyone who has had the chance to observe the Ba'Aka track gorillas in the forest would admit that there is some kind of magic to it. It is the ancient knowledge of hunters and gatherers, put into practise for something many of them view as complete madness; the habituation of gorillas![69]

While their reputation for tracking has given some of the disadvantaged Twa employment in the tourism sector, it has also caused some to become involved in violent conflict. During the civil war in the Democratic Republic of Congo from 1998 to 2002, often called 'Africa's World War' because many neighbouring countries were involved, Twa in the eastern Kivu region were forced to work as trackers and guides for Rwandan soldiers fighting local Mai Mai militias. After the departure of the Rwandan forces, the Mai Mai took revenge upon the Twa. During the same conflict Twa in CAR were slaughtered by Congolese rebels who wanted to exploit forest resources.[70]

It has long been known that excellent trackers exist among other African groups including those who historically practiced hunting and gathering, pastoralism, cultivation and some combination of all three. In the 1880s, a decade after his first trip to Southern Africa, Gillmore returned to hunt in the Kalahari area where he admired the tracking ability of the Bakalahadi. 'The skill of these men in tracking game I have previously noticed as something wonderful, far surpassing Indian red man or Oriental _chicaree_, and in this instance they afforded us a specimen of their superlative talent'.[71] Botswana safari industry pioneer Alec Campbell, while acknowledging the remarkable ability of the Bushmen, also considered the Bakalahadi as 'extremely good trackers'.[72] Shortly after the First World War, F. J.Bagshawe, a British district officer in Tanganyika failed to enlist Hadza hunter-gatherers as game trackers and believed that 'Naturally, they are expert hunters, and I doubt if better trackers are to be found in the world'.[73] During the early twentieth century an ethnographer in Northern Rhodesia reported that 'The Lambas are very good trackers of the spoor, and I have known some who could follow a trail almost as quickly as a Bushman'.[74] Just before the First World War, Paul Von Lettow-Vorbeck, German military commander in East Africa, indulged in hunting big game and often employed Maasai guides whose 'vision and skill as trackers are astonishing'.[75] Hunting rhinoceros and elephant on the plains of East Africa in the 1920s, the American Burbridge 'secured as trackers a couple of Masai who were very resourceful fellows'. Their leader 'hunted stark naked always, was an expert with the spear, a skilled tracker'.[76] John Alexander Hunter, a record setting big-game hunter in Kenya during the early and middle twentieth century, wrote that:

> In tracking, the assistance of a native tracker must be sought. The European is all right so long as he is following plainly defined spoor, but when the tracks fade away on hard ground it is only the keen eye of the native that can detect the signs that are invisible to you and me. Nothing is to be learnt here: it is a natural gift; and when I follow a tracker of the Walungulu or Wakamba tribes as he makes his bare-footed way tirelessly through thorny bush and pathless waste, pausing only occasionally to pull a thorn from the sole of his foot, I am filled with admiration and humility.[77]

Major G. H. Anderson, a British veteran of the East African campaign of the First World War and first president of the East African Professional Hunters'

Association in the 1920s, claimed that 'most Somalis are excellent trackers and seldom loose the spoor'.[78] For C. J. P. Ionides, a renowned British game warden and naturalist in Tanganyika who studied snakes, 'the two best trackers I ever had, both Kamba, started by following lizards and rodents' at age five.[79] Nick Steele, a game warden in South Africa's Natal province in the 1960s and 1970s, observed that 'The Zulu trackers were really impressive, pointing out a scuff of spoor here and there which would have gone unobserved by the rest of us'.[80] At the same time, the most famous tracker in the Natal Parks Board was Zulu game scout Magqubu Ntombela whose skill was refined during

> years of experience tracking animals, and humans too. From a tiny child Magqubu had learned this craft first taught by older herd abafana (boys) in following cattle or goats that got lost in the hills, then by his father, uncles, and cousins when hunting game.[81]

For British bush survival expert Ray Mears, who had been highly impressed by Namibia's Bushmen trackers, 'the best trackers I've ever come across were the Ovambo people in Namibia'.[82]

By the beginning of Africa's wars of decolonization in the 1950s, marginalized minorities were well known for tracking prowess and association with white hunters, and state security forces recruited them to find elusive guerrilla fighters. This was facilitated by the lack of contact that African nationalist movements had with remote and sometimes despised communities such as the Ndorobo in Kenya, Shangaan in Rhodesia and Bushmen in South West Africa. African nationalism had originated in larger population centres such as colonial towns and cities beginning with the small westernized African elite and spreading, particularly after around 1945, to the urban working class. With the turn to armed struggle in settler states such as Kenya and Rhodesia, nationalist movements took to the bush and began to recruit from large rural communities who had links to the urban population and who had grievances related to loss of ancestral land and the exploitive way they had been incorporated into the colonial capitalist economy. Since nationalism was related to western education and demands for economic development, supposedly primitive minority groups in isolated areas were usually left out of this process to some extent. Ironically, their limited absorption into the colonial economy made them ripe for recruitment by colonial security forces which also frequently played on historic tensions between marginalized minorities and larger neighbouring communities. Of course, the main problem with recruiting from minorities is that they could only supply a relatively small amount of manpower compared to the larger populations among whom the insurgents drew new members. As we will see, employing trackers from marginalized minorities celebrated for skill in this field became popular with state security forces in Africa's late twentieth century counter-insurgency campaigns but there were never enough of them to meet the demands of expanding guerrilla wars.

2 TRACKING AND COLONIAL WARFARE

The nineteenth century wars of colonial conquest in Africa are usually thought of as representing the victory of advanced European technology particularly involving quick-loading firearms against relatively primitive African groups armed with spears, clubs, shields and some obsolete guns. However, African man-power in terms of rank-in-file soldiers, supply carriers, labourers, wagon-drivers and scouts was central to colonial success. Within this context, the utilization of the indigenous hunting skill of bush tracking represented an important element of colonial warfare in Africa though it was not often thoroughly recorded by European officers who were usually the only literate witnesses to these conflicts and who might not have been eager to share their military glory.

In pre-colonial Africa, as elsewhere, weapons and skills used in hunting were often transferred to warfare. Tracking must have been part of this. It is generally assumed that as most Africans began to develop larger agricultural and pastoral settlements, hunting skills such as tracking declined but were retained by small groups of hunter-gatherers who became specialists. Increased demand for ivory and other hunting products, a result of expanding external trade during the eighteenth and early nineteenth centuries, meant that powerful Southern African states came to rely on the tracking abilities of small bands of hunter-gatherers. By the mid-1800s, in what is now northern Botswana and western Zimbabwe, the rival centralized states of the Ndebele and Ngwato, and sometimes European elephant hunters regularly employed Khoisan trackers. Around the same time, Khoisan hunters also became important suppliers of ivory to the nineteenth century Mpondo state along the Indian Ocean coast.[1] Nonetheless, and as illustrated in the previous chapter, tracking skills did not disappear entirely from settled African communities. For example, in pre-colonial times Zulu chiefs would demarcate a meeting place where trackers and warriors would gather to discuss an upcoming hunt and then depart with the former leading the way in search of game.[2] The lack of documentary records relating to pre-colonial African history and the limitations of oral traditions means that it is almost impossible to reconstruct the role of tracking in African warfare before the coming of Europeans.

Africa was not the only scene of colonial warfare where tracking became important. Nineteenth century North American literature celebrated frontier trackers such as James Fenimore Cooper's popular fictional character Natty Bumppo otherwise known as 'Hawkeye'; a white man raised by indigenous people to become a renowned warrior and hunter during the late eighteenth century. In its series of late nineteenth century campaigns against the indigenous communities of western North America, the United States Army employed auxiliary scouts who were usually indigenous or mixed-race men but also included a few whites with experience in the region such as the famous Jim Bridger and Kit Carson. The main job of such scouts was to collect intelligence. Since the indigenous people of the American west fought hit-and-run wars facilitated by the great mobility of their horse-culture, army scouts played a central role in locating enemy camps that often became the target of surprise attacks at dawn. As such, tracking (or trailing as it was often called in that time and place) was an indispensable skill for army scouts and it was claimed that indigenous trackers could follow a trail that most whites could not see. Given the importance of horses in these campaigns, tracking skills included the ability

> to estimate how long ago a party had passed by examining horse dung to see how far toward the center it had dried, and whether a party of Indians included women by the position of urine in relation to the horse's hoofs; women rode mares, while an all-male party on stallions was more likely to be a war party.[3]

While many white Americans believed that indigenous success in tracking was derived from innate and animal-like abilities and not intellect or reason, these scouting skills were developed throughout a lifetime as hunters and warriors. Furthermore, scouts were often important in combat, served as interpreters during negotiations and their very existence demoralized many indigenous people and convinced them that it was impossible to fight both the whites and their local allies. Despite their significance or perhaps because of it, indigenous scouts were usually ignored in official reports by white officers. However, according to historian Thomas Dunlay,

> Their reconnaissance and trailing function was indeed a prerequisite to any effective military action against hostile Indians. In addition, they extended the army's capabilities by making it possible for the soldiers to operate, to an extent, like Indians themselves.[4]

During the guerilla campaigns of the American Civil War (1861–5) both sides, perhaps learning from previous colonial wars, employed indigenous trackers. The Federal army used trackers from Kansas reservations in the west and the Confederates employed Cherokee trackers against Cherokee loyalist raiders in North Carolina.[5] Around the same time, British settler para-military forces in Australia such as the Native Mounted Police in Queensland engaged indigenous trackers in a long campaign to terrorize and dispossess indigenous communities as well as pursue white outlaws. During the late 1870s the famous Australian 'bushranger' or bandit Ned Kelly particularly feared the police black trackers sent after him. Although these para-military forces were disbanded in the 1890s following the complete subjugation of indigenous groups, indigenous trackers continued to be employed by Australian police forces until the late twentieth century. With the rise of modern indigenous political consciousness in Australia, the term 'black tracker' has become synonymous with 'traitor'.[6]

Along the eastern frontier of the Britain's Cape Colony at the southern tip of Africa, tracking became central to an aggressive colonial border raiding policy against the indigenous Xhosa. After supporting the Xhosa leader Ngqika in a dispute with his local rival Ndlambe, the British imposed the 'Spoor Law' in 1817 which empowered local white settlers to track allegedly stolen livestock across the border into independent Xhosa territory and repossess them by force. The tracking was often done by Khoikhoi who worked as herdsmen for white settlers or had enlisted as soldiers in the Cape Mounted Rifles (CMR) which was responsible for policing the frontier. Although the 'Spoor Law' was said to be based on an historic Xhosa method of dealing with stock theft by tracking, it was greatly abused and led to a long series of settler incursions against Xhosa communities. In late 1834 many Xhosa groups retaliated by attacking the colony which initiated the Cape-Xhosa War of 1834–5.[7]

Within the context of the 'Spoor Law', colonial forces in the Eastern Cape became frustrated with the constant pursuit of trails of supposedly stolen livestock. In 1825 the commandant of the Cape's eastern frontier was instructed that one troop of Cape cavalry

> is to protect the whole of this country of the Baviaan's River, by sending out such patrols as may be required, and attending most promptly to all applications and reports of Caffre (Xhosa) depredations, and following the spoor of the depradators [sic].[8]

There is evidence that Xhosa and other cattle raiders made use of what are now referred to as anti-tracking techniques to elude colonial patrols. Colonel Henry Somerset, the CMR commander, wrote that 'I am fully aware of the cunning of the Caffres, and the artifice made use by them to mislead my patrols when in pursuit of depredators, and their endeavours to throw the blame on the peaceable kraals'. Somerset instructed his patrols to follow a trail until they lost it, mark the spot and seek assistance from the local Xhosa chief and failing that to report to the nearest colonial post.[9] In December 1829 a group of Bushmen (Khoisan) who had stolen cattle from Boers near the colonial settlement of Graaf Reinet poisoned a water source which forced a Boer commando that was tracking them to turn back because of lack of water.[10] An 1834 account by a local Boer official illustrates the frustrations of tracking stolen livestock:

I followed the spoor as far as the Blink Water, where it was obliterated: the wind was so strong that I could not keep the trace any further. The following day I tried again to find the spoor on the other side of the Kat River, but did not succeed; on my return home on the 29th, I found the Caffres had again stolen seven oxen; I found the place where they drove them together, but rain fell, and the spoor was obliterated.[11]

In the early 1840s, after the 'Spoor Law' had been superseded by new frontier arrangements, settlers in the Eastern Cape reported that Xhosa cattle thieves 'had become so adroit in evading detection, by defacing the spoor or traces of the cattle over the several passes into Kafirland, that most of the cattle and horses carried over the boundary were irretrievably lost'.[12] While it is possible in some cases that African anti-tracking skill was exaggerated by colonial officials eager for an excuse to raid livestock, the decades of livestock raiding back and forth across the frontier must have produced some excellent cattle thieves.

The large nineteenth century European led armies that invaded African territory usually employed local African, mixed-race or white settler scouts to move ahead of the main body and collect information. This often involved tracking. When forming a commando for raiding or defence, Boers in the Cape usually brought along Khoikhoi servants called 'after riders' who performed many support functions such as caring for horses, standing sentry, cooking food and working as scouts and trackers. While many scouts for colonial armies were civilians employed on an ad hoc basis, some locally recruited standing military units became central to these activities. In the Cape Colony, the CMR was formed in the early 1800s with European officers and Khoikhoi and mixed-race soldiers, and from the 1850s was phased out in favour of the Frontier Armed and Mounted Police (FAMP) which consisted entirely of white settlers. Indigenous auxiliary units such as Khoikhoi infantry battalions and Fingo (or Mfengu) levies were formed during specific wars and then disbanded when hostilities ended. All these African forces supplied trackers to the British army in the Cape Colony. While it might be tempting to see the colonial recruitment of Khoikhoi soldiers as related to this group's reputation as possessing excellent trackers, they were the first indigenous people to come under colonial rule in what is now South Africa and, therefore, were the first to be formed into a colonial military unit. That said, the Khoikhoi riflemen did earn a high degree of admiration for their tracking abilities. In 1834 a British officer reported that the trickery of the Xhosa cattle thieves was 'only equalled by the perseverance and wonderful sagacity of the pursuer, the Hottentot (Khoikhoi) soldier, who will follow on the "spoor" for days together until he comes up with and recovers the booty'.[13] A British officer who served in the Cape-Xhosa War of 1834–35, in the context of pursuing Xhosa raiders with a CMR patrol, wrote that

It was interesting to see the sagacity of the Hottentots in looking for the spoor. The grass was examined to see if it had been pressed down by the foot; and if twigs had been broken off the bushes; and stones were carefully lifted to note if they had been moved.[14]

The same officer remarked that the Khoikhoi soldiers 'rival North American Indians in tracking an enemy by his marks, though several days old, on the ground and on the bushes'.[15] In the wake of that conflict, British missionary William Shaw recommended the expansion of the CMR especially because of its soldiers' tracking skill. In pursuing cattle stolen by the Xhosa, Shaw claimed, 'The Hottentots of the colony connected to the Cape Corps, some of whom should always be of the party, perfectly understand the mode of following up the spoor'.[16] Trackers from other African groups were also sometimes employed such as 'a Bechuana (Tswana) well versed in spoor' who worked for a British officer fighting in the Cape-Xhosa War of 1835.[17] After the conquest of the western Xhosa who became colonial subjects within the new territory of British Kaffraria in 1848, the British created a 'Kaffir Police' (Xhosa Police) force which played a major role in tracking stolen livestock. For George McKinnon, chief commissioner of the new colonial territory, these 'Kaffir Police' were far superior for this type of work as:

> the British soldier ... does not possess the faculty of tracking livestock ... nor will any length of service or experience in the country enable him to acquire this. The Hottentot of the Cape Corps, on the other hand, is well able to follow the traces of livestock, and to spoor (track) the footsteps of the Kafirs who steel it; but there is antagonism between the Kafir and Hottentot races; hence to revert to the system formerly pursued of sending out armed parties of this corps to make reprisals on the Kafirs for the thefts which they commit would ... lead to a war-cry.[18]

While the British in the Cape learned that depending on subject African communities for military manpower could be risky, they still needed skilled trackers with local experience. Colonial racial oppression against the Khoikhoi and the increasing British reliance on more numerous Xhosa-speaking Fingo allies prompted rebellion by the Khoikhoi Kat River Settlement in the Eastern Cape and a mutiny in the CMR at the start of the Cape-Xhosa War of 1850–53. After Khoikhoi and Xhosa resistance had been crushed, Governor Sir Harry Smith believed that the Khoikhoi troops had to be gradually replaced by 400 young British immigrants who 'would soon acquire the local habits which are requisite, those of tracing the footsteps, or "spoor", as it is here termed, of men and cattle'.[19] Around the same time another British officer who had served in the Cape during the 1830s recommended to the British government that the Cape forces discontinue the use of large Khoikhoi units but retain 'a few Hottentots who should accompany the Europeans to act as guides, from their knowledge of the spoor and the habits of the Kafirs'.[20] In the late 1870s a British colonial official described the members of the Cape's FAMP, the exclusively white unit that replaced the CMR, as 'all colonists, hardy, robust, grown men, used to the rifle and the saddle from boyhood – men who could "spoor" a lost ox, day and night, with eyes like a hawk and ears like a hare'.[21] However, in the 1860s former British policeman and FAMP sub-inspector Edward Wilson admitted his reliance on African trackers and pointed out that

It is not a very easy matter to follow up a spoor. Occasionally it is tedious and very critical work. The Natives are decidedly the best adepts at it; this, I may remark, being their forte. For other public purposes they are useless.[22]

Although the FAMP had no official African members, it employed 'a few Native Foreigners' as interpreters and for 'following the spoor of stolen cattle'.[23]

In many of these colonial campaigns, following the obvious signs left by large bodies of African warriors or herds of cattle was not difficult for most trackers though sometimes resulted in dramatic success. In early 1835, after Xhosa raiders had attacked colonial settlements and used the thick bush around the Fish River to elude columns sent after them, a strong patrol of British infantry and mounted Boers 'following the Kaffir spoor, came upon a body of the enemy in a ravine near Commatty's Drift; slew some, and dispersed the rest: capturing one hundred head of cattle'.[24] In early June 1846 a Cape colonial cavalry patrol under Somerset ventured out from Fort Peddie to distract the surrounding Xhosa from a train of empty supply wagons that had started its journey back to the regional capital of Grahamstown. Somerset's Khoikhoi and British horsemen found and followed the tracks of a large group of Xhosa which ultimately led to the interception and defeat of Siyolo's unsuspecting warriors at the decisive Battle of the Gwangqa River.[25] According to a colonial soldier:

> As daylight broke we, for I was with the column, came upon the track or spoor of this body of Kaffirs at right angles with our own march, who must have passed over the open just before us. The trace showed a broad space of about twenty yards wide, with the grass trodden down and the dew dispersed from it. The General at once followed up this 'spoor', and as the sun rose we came suddenly upon a large mass of the enemy who had fires lit, and were at their morning repast of dried flesh and parched Indian corn.[26]

The discovery of diamonds in the Northern Cape in the late 1860s and gold in the Transvaal in the 1880s led to the dramatic acceleration of colonial conquest in southern Africa. Trackers were involved in these campaigns. When the British army in the Colony of Natal invaded the neighbouring Zulu Kingdom in 1879, it deployed African scouts who searched for signs of enemy movement and found that Zulu scouts on horseback had observed their movements.[27] During the same war, a colonial officer reported that a patrol of six mounted European and African scouts would be

> guided by the more keen instinct of the natives in tracing the spoor of the enemy and discovering all such information as they might be able to obtain on what had happened on the previous night.[28]

Once superior colonial firepower had defeated an African force on the battlefield, tracking was particularly useful in finalizing the victory by pursuing fleeing leaders and their entourages. After the final defeat of the Zulu at the Battle of Ulundi in 1879, the British dispatched patrols of colonial cavalry and Natal Native Contingent, local African allies, to track the fleeing Zulu King Cetshwayo who was eventually captured and temporarily exiled to Britain. The former Zulu Kingdom, one of the region's greatest African powers, was then broken up into 13 separate chiefdoms and endured an horrific civil war in the 1880s.[29] In what is now western Zimbabwe, Ndebele King Lobengula fled north toward the Zambezi River after his armies' 1893 defeat by the conquering forces of Cecil Rhodes' British South Africa Company (BSAC). The subsequent chase was described by Frederick Russell Burnham, an American frontiersman who had joined the invading army's scouting element and became one of the few survivors of the famous massacre of the Shangani Patrol which was eventually ambushed and annihilated by the Ndebele:

> Under Major Allan Wilson, the best trackers of Africa – Hottentots and Masarwas (Bushmen) from the desert, whose existence largely depends on their ability to follow an antelope or recognize the faintest trace of small game – were put to work to discover the trial of the king's wagons. We had also in our force others especially trained in trailing: Boers of the Transvaal, Australians, and frontiersmen from all parts of the world. The final tracking, however, depended upon Ingram, an American, Bain, a Canadian, and myself. We found that we could out track the black scout even at his best. Many times we lost the spoor, but we always found it again.[30]

As this suggests, experienced trackers from North America's and Australia's fading frontier zones had made their way to Southern Africa to ply their trade. After the ill-fated Wilson column encountered the Ndebele and began to retreat, two mounted scouts were sent to fetch reinforcements from Rhodes at Lobengula's former capital of Bulawayo which had been turned into a colonial headquarters. Burnham became concerned when he noticed that the hoof prints of the scouts' tired horses were superimposed by Ndebele footprints which meant that the messengers were being closely pursued by about 10 or 15 of the enemy. Boer frontiersman Johan Colenbrander, however, carefully analysed the spoor and concluded that the pursuers were young Ndebele men who he believed lacked determination and would abandon the chase. Since the two messengers always backtracked and concealed themselves in the bush before resting, they noticed the Ndebele warriors tracking them and were able to escape. According to Burnham, who seemed dismissive of African tracking skills, 'This simple ruse would not have deceived an American Indian, but it was sufficient to enable the scouts to elude the Matabele'.[31] Further colonial pursuit of Lobengula became unnecessary as he died of illness shortly after the demise of the Wilson patrol. The subjugated Ndebele Kingdom was then incorporated into Rhodes' new white settler colony of Southern Rhodesia (today's Zimbabwe).

Tracking became most important and more difficult in conflicts where African groups such as the Xhosa of the South Africa's Eastern Cape resorted to hit-and-run bush warfare to minimize the advantages of colonial firepower and mobility. A British colonial official leading a patrol of Khoikhoi troops near the Amatola Mountains during the Cape-Xhosa War of 1835, wrote that 'two or three of the enemy's scouts were observed watching our movements. And having ascertained by their spoor the direction they had taken, we continued following them'.[32] Describing the danger of pursuing an elusive foe who could easily stage an ambush, a British officer fighting in the same conflict wrote that 'It is certainly no child's play tracking through the dense bush by a narrow path in Indian file, having a volley of musquetry (sic) suddenly poured on the party from above'.[33] Lieutenant Moultrie of the British 75th Regiment, based at Fort Peddie in September 1835, described an operation which illustrates the importance of Boer and African allies in tracking the hostile Xhosa:

> Having observed their spoor (traces) in the Fish River, in different small bodies, and hearing also of several robberies which had been committed in the colony; I yesterday again went into the bush for the purpose of intercepting them. I lay in wait in one of their most favourite paths; and during the night I succeeded in killing three of them. Shortly after my return to the camp this morning, a man arrived from Jan Tzatzoe, to inform me that the spoor of a number of Kaffirs had been seen by some of the women in the direction of the Guanga (sic), where the cattle of the tribe usually graze ... I nevertheless ordered Field-cornet Piet Uys, with two and twenty of the burghers, to proceed in the direction alluded to; and if the information was correct, to endeavour to cut them off. It turned out to be true, that a body of fifty or sixty Kaffirs had come through; but upon finding themselves discovered, they had retreated towards the Keiskamma. The field-cornet followed them; and upon being joined by Tzatzoe and his men, succeeded in surrounding them in a blind (dry) river near the Line Drift ... eighteen of them were killed and a number were wounded.[34]

Jan Tzatzoe was a Christian Xhosa chief who sided with the British during some of these wars, and supplied them with trackers and information on enemy tracks. Leading a colonial sweep by British troops and Fingo irregulars of the thickly forested and rugged Waterkloof Highlands during the Cape-Xhosa War of 1850–53, Lieutenant Colonel Thomas Fordyce reported that he had encountered no opposition but 'It appeared, however, by the fresh spoor and quantities of chewed roots which we passed, that a large number of Kafirs must have passed up very recently'.[35] Soon after writing this, Fordyce was involved in a skirmish in the Waterkloof where he was killed by a sniper who was probably a Khoikhoi mutineer from the CMR. Tracking was important to colonial mopping up operations in the same war as in November 1852 it was reported that:

> Major Horne, 12th Regiment, with a patrol from Governor's Kop, follows up the spoor of a party of Kafirs into a dense kloof of the Botha Hill range, where Lieutenant Goodison, with thirteen of the Cape Mounted Rifles, assails the enemy, kills nine Kafirs whilst Ensign Adams, 12th Regiment, with a small detachment from Botha's Hill post, cuts off the retreat of the fugitives, killing three of their number.[36]

During the Cape-Xhosa War of 1877–78, the last instance of resistance by the hard-fighting Rharhabe and Gcaleka Xhosa, the British employed Fingo allies who were sent into the thick Pirie Bush to track Xhosa rebels and cut off their supply routes.[37] A British officer who fought in this conflict wrote that

> you may be surrounded by a crowd of (warriors) in the bush, and unless you have come across their spoor you may be ignorant of their proximity till, with a rush, a red form with a quivering assegai appears within a few yards of you.[38]

Avoiding the pitched battles that had characterized the Anglo-Ndebele War of 1893–94, the Ndebele who rebelled against BSAC rule in 1896–97 used the rocky Matopos Hills as a sanctuary. The account of Lieutenant Webb, sent out with a mounted patrol in April 1896 to investigate the burning of a European owned store by Ndebele rebels, illustrates how awareness of tracking influenced military actions.

> The place we found had been burnt to the ground by the Matabele during our stay at Gwanda, and judging by the spoor a large number of them had been at work. We decided to stay the rest of the day at this place, and were careful to take every necessary precaution in case of a night attack.[39]

Attempting to locate the rebels, British cavalry officer Robert Baden-Powell conducted solitary or very small reconnaissance patrols which he believed had less chance of being noticed than larger parties. He wrote that 'what I prefer is to go with my one nigger-boy, who can ride and spoor and can take charge of the horses while I am climbing about the rocks to get a view'.[40] Baden–Powell had already developed a fascination for tracking during previous service in India and Zululand, and had authored a manual for training cavalry which included an emphasis on the importance of this skill in collecting intelligence. He wrote that

> Wheel-marks and footmarks will always afford a deal of information as to the numbers, composition and direction of a force, and if you are practised noting foot-tracks you will be able to tell how long ago they were made, and consequently how far off the force is.[41]

Applying and refining his ideas in the Matopos, Baden-Powell wore rubber soled shoes which silently gripped the rocky ground, tried to move backwards over his own tracks to confuse the enemy, found that wiping away one's spoor on hard ground could easily be detected by rebel scouts and avoided enemy ambush by never returning to camp via the same route. His tracking and scouting partner, and probably mentor, was Jan Grootboom, a Fingo from the Eastern Cape who sometimes disguised himself as an Ndebele and infiltrated their settlements to gather information. Baden-Powell noticed that although the Ndebele tracked their enemies, they rarely tried to conceal their own spoor except by jumping over established trails.[42] Pioneering methods used in the region's counter-insurgency wars of almost a century later, Baden-Powell collected considerable intelligence on the rebels by analysing all the tracks found in a specific area including determining their age and later in the war led larger patrols that tracked and tried to engage Ndebele forces. This experience convinced him that a single reconnaissance scout, well-trained in 'the art of noticing smallest details, and of connecting their meaning, and thus gaining a knowledge of the ways and doings of your quarry', was not just important in colonial campaigns but remained indispensable in modern warfare between technologically advanced opponents.[43] In this emerging view, tracking was no longer an innate ability of supposedly primitive indigenous people but a science that Europeans could master by observation and deduction. As the British Empire expanded during the 1890s, military scouting would be divided into two categories; the purist trackers, and the rough and tumble skirmishers. Baden-Powell seemed to represent the former.[44] It was during the Ndebele Rebellion that Baden-Powell met the American scout Burnham, chief of scouts for British forces in the campaign, from whom he learned to incorporate tracking into a broader array of wilderness knowledge and survival skills called 'woodcraft'.[45]

Tracking became an essential skill during the South African War or Second Anglo-Boer War of 1899–1902 and was associated primarily with the Boer commandos which often consisted of frontier farmers and hunters. On one occasion, the young Boer fighter Deneys Reitz became separated from his colleagues and used tracking to find them.

> I saw several fresh hoofmarks on the ground. On examining these, I recognized the slightly malformed marks of Michael du Preez's pony, and closer investigation showed me the footprints of men which I knew at once as those of some, if not all, of my seven missing companions... I lost no time in following their spoor.[46]

Of course, by this time not all Boers were competent trackers. While the Transvaal Boers who fought the British in the First Anglo-Boer War of 1880–1 were accomplished in hunting skills such as tracking and shooting, by the 1890s these abilities had begun to fade among the Boers as wildlife had been wiped out in many areas and the growth of a mining economy had prompted urbanization.[47] Early in the conflict, at the personal request of British commander Frederick Sleigh Roberts, Burnham left Alaska where he was prospecting for gold and returned to Southern Africa to take up the position of 'chief of scouts' for the entire British army in the region.[48] The prolonged Boer sieges of British garrisons at Ladysmith, Kimberley and Mafeking during the first months of the war were the scenes of tracking and anti-tracking. In March 1900, during the Boer siege of Mafeking where Baden-Powell commanded the hungry British garrison, a cattle raiding party of 25 African auxiliaries realized that it was being tracked by a Boer patrol, doubled back on its spoor and ambushed the pursuers killing eight and wounding seven.[49] The South African War began as a large conventional conflict but with the eventual British occupation of the Boer republics, some Boer commandos refused to surrender and embarked on a guerrilla struggle which relied on support from Boer civilians and thorough knowledge of the environment. In response, the British military recruited Boer prisoners who were formed into the Transvaal National Scouts and Orange River Colony Volunteers. These new units used Boer tactics such as muffling horse hooves and rifles with cloth so they would make less noise at night and undertook tracking. Black men who had served as 'after-riders' in the Boer commandos and then deserted or were captured were also recruited by the British as scouts and British commanders much admired their tracking abilities. In the frontier regions of the northern and north-western Cape, the British mobilized the tracking and other bush skills of mixed-race men by enlisting them into the Namaqualand Border Scouts, Bushmanland Borderers and other smaller auxiliary units.[50] As evident in Southern Rhodesia during the 1890s, the British fighting the Boers in South Africa also mobilized frontiersmen from other parts of their empire and it was thought that those from 'Canada and Australia were very good trackers'.[51] Canadian Sam Steele, a pioneering officer of the North-West Mounted Police who had participated in the suppression of the 1885 Metis rebellion in Manitoba, was sent to South Africa in command of a cavalry reconnaissance unit called Lord Strathcona's Horse made up of western Canadian frontiersmen. After the war he remained with the occupation forces and commanded a division of the South African Constabulary until 1906. One of Steele's biographers maintains that 'Sam's skills as a scout and tracker, and his experience fighting a crafty adversary in thick bush, made him the obvious' choice to command these units that specialized in pursuing Boer guerrillas.[52] Despite their formal exclusion from the all-white Australian military, some indigenous Australians served in the Australian forces in South Africa and may have worked as trackers as many did for the police back home. In January 1902 British commander Lord Horatio Kitchener, frustrated by continued Boer guerrilla attacks, requested that the new Australian federal government send Aboriginal trackers for service in South Africa and four were sent to the Bloemfontein police. Though it has been contested, some historians maintain that 50 Australian Aboriginal trackers were abandoned in South Africa after the war because non-whites were barred from entering Australia.[53]

The experience of tracking in colonial wars did not just influence the development of the British Army during the early twentieth century but broader imperial society as well. Charles E. Callwell, a British officer who wrote a highly influential book on small wars during the late nineteenth century, advised that

> scouts are generally natives who cannot be trusted far out of sight; but Europeans who have long lived an open air life in a theatre of guerrilla warfare, who are accustomed to track footprints and who are adept at the hunter's craft, will move miles ahead of the fighting force and can sometimes fix the quarry at several marches distance.[54]

Baden-Powell, who won fame by commanding the besieged Mafeking garrison during the South African War, continued to write memoirs and training manuals which popularized tracking as part of military reconnaissance. Ultimately, tracking also formed a major component of Baden-Powell's civilian 'Boy Scout' movement which promoted vigorous masculinity and frontier skills among British youth, and eventually spread around the world. His military oriented *Aids to Scouting for N.C.O.'s and Men*, first published in 1899 and reprinted many times, was rewritten for children and published as *Scouting for Boys* in 1908 and became the central text of the Scout movement. Both books contained detailed chapters on the mechanics of tracking, included training games and examples of the author's own experiences in Africa, extolled the tracking and wilderness skills of American frontiersman Burnham who became a role model for the movement, and encouraged readers to emulate the observation and deductive abilities of fictional detective Sherlock Holmes.[55] Perhaps as a nod to the role of Africans in colonial hunting and warfare, the entirely white Boy Scout movement in early twentieth century South Africa reacted to calls for racial integration by establishing an alternative black group called the 'Trackers' which failed to attract recruits who wanted inclusion in the main association.[56] A less well-known and much less enduring British imperial para-military group to emerge in the wake of the South African War was the Legion of Frontiersmen which sought to organize those with frontier skills such as tracking as a pool for military recruitment in Britain and the dominions. Like the Boy Scouts, the Legion also had an official manual. In 1911 Legion founder Roger Pocock, a British self-styled adventurer who had fought with the North-West Mounted Police against the Metis in the 1880s and the National Scouts during the South African War, published *The Frontiersman's Pocket Book* which covered subjects such as scouting, horsemanship and demolitions.[57] In the aftermath of the South African War, and with the influence of Baden-Powell, the British army mandated that each cavalry regiment establish a small reconnaissance element with personnel trained in navigation, observation and tracking.[58]

Tracking continued as an important feature of warfare and internal security in early twentieth century colonial Africa. Trackers played a central role in the German East Africa campaign of the First World War where, after a disastrous British attempt to invade from Kenya in 1914, German commander Paul von Lettow-Vorbeck staged hit-and-run attacks across the border into British ruled Kenya and eventually fought a long series of delaying actions against overwhelming Allied invasion forces. Both sides formed irregular scouting and intelligence units often made up of mounted white settlers with hunting experience or African peoples with martial reputations. Among these was the Nandi 'Skin Corps', so called because of their disdain for clothing, led by British officer Major J. Drought. In Kenya, during the early days of campaign, British aristocrat Berkeley Cole formed an irregular cavalry unit consisting of Somalis who patrolled the strategically important Uganda railway and the border with German East Africa. Celebrated white hunter Denys Finch Hatton joined Cole's Scouts and observed that the Somalis 'turned out to be brilliant trackers, as Berkeley had surmised'.[59] Australian Charles Joseph Ross, who had led a small unit of Canadian scouts during the South African War and arrived in East Africa in 1904 to hunt elephants, led scouts that patrolled the Kenya border during the early days of the First World War. Ross was an experienced tracker having learned the skill while living with North American indigenous people and then worked as a scout for both the United States and Canadian armies in wars of conquest against such communities in the late nineteenth century.[60] The British tried to use scent tracking dogs to follow German raiders attempting to sabotage the Uganda railway but found that the sandy, dry ground did not retain enough scent. Kamba trackers had more success.[61] Indeed, a South African Boer tracker facilitated one of the most strategically important actions of the East Africa campaign. In 1915 Major P. J. Pretorius, a South African who had hunted in Central and East Africa since the 1890s and who had alienated German authorities by poaching elephant, used bush skills and disguises to locate the German warship Konigsberg which had threatened British shipping in the Indian Ocean and was now hiding in the Rufigi River Delta. Familiar with the area, Pretorius collected technical information on the river and its tides which proved instrumental in the destruction of the Konigsberg by shallow-bottomed British gun-boats. Eventually, Jan Smuts, the South African commander of British imperial forces in East Africa, appointed Pretorius as 'chief scout' and sent him on long missions in the bush gathering information on the elusive German columns.[62] Although the German East Africa campaign cannot be called an insurgency, Pretorius favoured several practises which would become standard in later counter-insurgency theory. He cultivated friendly relations with rural African communities to gain their cooperation and deny the same to the Germans, and he often recruited African trackers and scouts from among captured German colonial troops. According to Pretorius, his hunting and tracking skills served him well during the campaign:

> I was still following jungle trails, still the hunter, only the quarries were humans instead of animals. Men left their spoor as did animals, and the practised bushman could read the same story in the ashes of deserted camps as in the broken twigs and bent grass caused by the passing of a wild beast. One knew which way the enemy had gone, what was his strength, and how long since he had departed.[63]

Immediately after the war Pretorius was hired by the Cape provincial administration in South Africa to exterminate the elephants of the Addo Bush who were menacing nearby citrus farmers but after killing 100 of them he convinced authorities to preserve a remnant of the herd and the area eventually became a nature reserve.[64] The head of intelligence for British forces in East Africa was Major Richard Meinertzhagen, a veteran of British punitive campaigns against the Nandi a decade earlier and an avid naturalist, who was 'a very experienced hunter and tracker, and he was utterly self-sufficient in the bush if he had to be'.[65] Specifically raised for the East Africa campaign by Britain's Legion of Frontiersmen, the 25th Royal Fusiliers had a number of men with hunting and tracking experience including Frederick Selous, the famous 'great white hunter' now in his 60s, who was killed by a German sniper in January 1917.[66] The Germans also employed hunters and trackers, and Lettow was kept informed of Allied movements by his own 'chief scout' and 'able companion', a Boer called Piet Nieuwenhuizen who had moved to East Africa via Rhodesia in 1906 and knew the area very well.[67] As a young officer in late nineteenth century Germany, Lettow had hunted to foster useful military skills such as tracking, and his 1904 experience fighting Herero rebels in German South West Africa, where he learned much from his Boer advisor and African auxiliaries reinforced his belief in the importance of tracking and marksmanship.[68] An Australian Boer War veteran, cattleman and politician, Arnold Wienholt had learned tracking from indigenous Australians and Bushmen in Angola where he had hunted lion just before the war. In November 1915, in what is now western Zambia, he led a small group of Rhodesian police that tracked and apprehended German soldiers and Boer rebels who had fled German South West Africa and were on their way to join their comrades in East Africa. Wienholt then enlisted as an intelligence scout with British forces in East Africa where he led numerous scouting missions during that campaign and was himself tracked and captured by German scouts including Nieuwenhuizen.[69]

During the early part of the First World War the British employed Bushmen scouts from Bechuanaland to keep track of German movements along the border with German South West Africa which would soon be invaded by South African forces.[70] Shortly after the war, police in South African administered South West Africa broke up Bushmen communities but also employed Bushmen trackers to apprehend other Africans deserting coercive labour policies. One such Bushman tracker was Native Sergeant Saul who was instrumental in pursuing the murderer of a Gobabis magistrate in 1922.[71] By the late 1940s, the police in South West Africa were trying to befriend Bushmen groups by treating them well and giving them small amounts of tobacco and salt. A magistrate advised district officers that 'If afforded proper and systematic treatment, experience has taught us that they can be very useful in the detection of crime. They are almost infallible as trackers'.[72] This relationship between the South African police and Bushmen trackers in South West Africa would continue for many years and inform early counter-insurgency efforts in that territory in the late 1960s.

During the nineteenth century European conquest of Africa, mostly black and some white trackers collected intelligence by moving ahead of large invading colonial armies and pursued defeated African rulers. Foreshadowing events in the second half of the twentieth century, trackers were at their most important to colonial forces when pursuing elusive hit-and-run African and Boer fighters.

At times special colonial military units were created such as the CMR, FAMP and Transvaal National Scouts that sought to mobilize local tracking and scouting skills. While trackers from minorities with a reputation for this skill such as the Khoikhoi and Bushmen were employed by colonial forces, many were from large groups and some were local white settlers or frontiersmen from North America and Australia. These colonial wars also created tracking celebrities such as Burnham and Baden-Powell who popularized the skill within military and social contexts. The importance of tracking in colonial conquest shows that possession of advanced military technology by colonial armies was not the only factor that led to the defeat of African groups. In colonial warfare, tracking usually represented a form of indigenous knowledge that was co-opted by colonial forces and used in conjunction with other factors such as superior firepower to suppress indigenous resistance. Although the horrific industrialized warfare that occurred in Europe during the two world wars certainly took the spotlight off tracking and other wilderness skills in Western militaries, a similar process of incorporating indigenous tracking skills occurred during the counter-insurgency campaigns fought in East and Southern Africa during the late twentieth century.

3 KENYA, 1952–6

In 1895 the British government took over the East Africa Protectorate, renamed Kenya in 1920, from the failed Imperial British East Africa Company that had originally attempted to colonize the area a few years earlier. Initially, Britain's main interest in the territory was as a railway corridor through which to extract resources from the agriculturally rich colony of Uganda in the interior to the Indian Ocean coast. Resistance from the coastal Swahili during the 1890s and the pastoral Nandi of the interior during the early 1900s was crushed. The first white settlers arrived in 1902 in a scheme meant to finance the construction of a railway to the agriculturally rich colony of Uganda. White commercial farming developed supported by a system of cheap black labour created by the imposition of taxation, the confinement of people in overcrowded reserves, a ban on blacks growing crops favoured by the settlers, a pass system which controlled black workers' movement and state labour conscription. During the colonial era the Kikuyu of central Kenya, compared to other people in the colony, experienced disproportionate dispossession as their fertile land which also had a healthy climate was favoured by European settlers and became the 'White Highlands'. Numbering around 30 000 in the 1930s and 80 000 in the 1950s, the small white settler minority developed an elite identity based on mainly British upper class origins and became politically influential in the running of the colony through their domination of the local legislative council. While Kenya's white settlers dreamed of becoming an autonomous dominion within the British Empire like Canada or Australia, they never gained any official powers of self-government. From the beginning of the colonial era, Kenya's security fell to a local police force consisting mostly of Africans and some Asians under European leaders and the King's African Rifles (KAR) a primarily African infantry force under British officers and some NCOs which, in peacetime, came under the authority of the Colonial Office. Settler political ambitions led to the 1928 creation of the Kenya Defence Force in which all young white men were to undergo military training and in 1937, within the context of rising international tensions that would led to the Second World War, this was morphed into a reserve infantry battalion called the Kenya Regiment which aimed to provide white junior leaders to the KAR during wartime and to respond to internal disturbances.[1]

Landless Kikuyu became squatters on European commercial farms where they served as compulsory cheap labour and during the Second World War many flocked to the growing city of Nairobi to seek better employment. At the same time, a relatively prosperous Kikuyu peasantry emerged that became loyal to the colonial state. Given early exposure to mission education and their disadvantaged situation, the Kikuyu people were instrumental in establishing the first African western-style political organizations in Kenya such as the East Africa Association and the Kikuyu Central Association of the 1920s which attempted to voice African grievances within the existing system. During the 1940s and early 1950s the moderate Kenya African Union (KAU) urged the colonial administration to eliminate racially discriminatory legislation and increase the tiny African representation in the colony's legislative council. The failure of this political campaign shifted the political momentum to more militant African leaders within trade unions and among the squatters in the White Highlands. Explanations of the subsequent Mau Mau insurgency range from the British colonial view that the predominantly Kikuyu rebels were suffering from a psychological disorder caused by overly rapid westernization to a Kenyan nationalist view that they were fighting for Kenyan independence to more specialized academic studies that see it as a narrow ethnic uprising caused by the imposition of a colonial settler economy.[2]

Although there had been intra-Kikuyu violence since the 1940s, the killing of a European woman and a Kikuyu chief in October 1952 prompted newly arrived Governor Sir Evelyn Baring to declare a state of emergency. The next day security forces launched Operation Jock Scott which involved the arrest of 180 suspected insurgent leaders in Nairobi including KAU president Jomo Kenyatta who was later convicted of treason. Militant leaders fled to the high forests from where they organized a series of violent attacks over the next few weeks. At the emergency's start British forces in Kenya consisted of 7000 men including 39 Brigade which consisted of three British infantry battalions flown in from Egypt, 70 Brigade comprising five KAR battalions (three from Kenya and one from each of Uganda and Tanganyika), and white settlers formed into an armoured car squadron and the Kenya Regiment. Within several months the arrival of Britain's 49 Brigade with two infantry battalions and an engineer regiment increased this force to 10 000 soldiers eventually assisted by an expanded Kenya Police of 21 000 which included recruits from Britain and a new 25 000 strong Kikuyu 'Home Guard'. Most of the British soldiers and police who arrived in Kenya, including many conscripts, had no experience in East Africa let alone in living or fighting in high altitude forests. This meant that Kenyan whites, with their local knowledge and expertise, influenced the course of the counter-insurgency campaign.

With a decentralized leadership, the Kenya Land Freedom Army formed small units in the forests of the Aberdares Mountains and around Mount Kenya, and organized a passive support wing in settled Kikuyu reserves. The British called the insurgent movement Mau Mau though the origin of the term is unclear. With around 12 000 insurgents of whom only 1200 were armed with rifles, Mau Mau was divided into three zones: the Central and Northern Aberdare Mountains under Dedan Kimathi, the Southern Aberdares led by Stanley Mathenge and Mount Kenya commanded by Waruhiu Itote called 'General China'. While Mathenge and Itote were both veterans of the Second World War, it was rare for other African veterans to join the rebels who generally lacked military experience. Although Mau Mau leaders attempted to establish a coordinating structure in August 1953, the insurgency consisted of a series of independent and sometime hostile local groups. Mau Mau lacked an ideology and theory of revolutionary warfare, most members were ill-literate or semi-literate, there was no external sponsor or cross-border staging area, and weapons were mostly captured or homemade firearms, spears and machetes. Insurgent recruits were bound to the movement by ritual oaths, and torture and mutilation of victims spread terror. Rebels were supplied with food and intelligence by local sympathizers, groups communicated by 'letter boxes' hidden in trees or under rocks and they became skilled at moving quickly and covertly through the bush. Unlike the anti-colonial insurgencies in parts of Asia during the 1950s and those in Africa during the next decade, the absence of Eastern Bloc footholds in colonial Sub-Saharan Africa at that time meant that the global Cold War did not superimpose itself on the Mau Mau uprising which remained a local affair.

The Mau Mau war was largely fought in central Kenya, about 100 to 150 kilometres north of Nairobi in the areas around the Aberbares Mountains in the east and Mount Kenya in the west between which was the 'White Highlands'. With an average elevation of around 3500 meters and about 100 kilometers long from north to south, the Aberdares consists of three ascending environmental belts; rainforest, bamboo forest and rolling moorland. The area is also characterized by deep forested ravines, rivers and waterfalls, and regular rainfall. Mount Kenya is the highest mountain in Kenya and the second highest in Africa with several peaks that are almost 5200 meters high. Around the base of the mountain is cool, fertile farmland, and above that are ascending rings of mountain forest, bamboo forest, and low timberline forest and shrub land topped by rocky, glacial peaks. Much of Mount Kenya and the Aberdares became national parks in 1949 and 1950, respectively, which meant the forest areas were mostly uninhabited and patrolled by a few state officials.

In June 1953 General George Erskine took over British operations in Kenya and created mobile units that attempted to clear insurgents from specific areas which were then patrolled by loyalist forces and police. Penetrating Mau Mau territory, the British had cut five separate seven-kilometre long tracks into the forest by August 1954 and positioned a battalion base at the end of each and as many as 20 more were under construction by the Royal Engineers.Launched in

late April 1954, Operation Anvil aimed to deprive the insurgents of their source
of supplies and recruits in Nairobi which was cordoned off and searched by 25
000 soldiers and police. All Africans were detained in barbed wire enclosures
until their identity was confirmed. Those from ethnic groups not associated
with the rebellion were released while Kikuyu, Embu and Meru were held. Some
20 000 men were moved to a detention camp for further screening and 30 000
women and children were evicted to rural reserves. From June 1953 to October 1955 the Royal Air Force (RAF) flew reconnaissance, propaganda leaflet
dropping and bombing missions over Kenya. Although RAF bombing was often
chaotic, it killed around 900 insurgents and compelled some groups to disband
and others to flee from the forests to the reserves.

Through a system they called the 'Pipeline', British officials colour coded
Kenyan prisoners white, grey or black. The 'whites' were the most cooperative
detainees who were returned to the reserves, the 'greys' confessed to having
taken the Mau Mau oath but were cooperative so they were moved 'down' the
'Pipeline' to local labour camps before eventual release, and the 'blacks' were
the most uncooperative prisoners who were sent 'up' the 'Pipeline' to special
detention camps. In the year after Operation Anvil the British had little success
in obtaining confessions from detainees as the camps were unprepared for the
massive number of prisoners. However, by 1955 the 'Pipeline' had become more
effective by moving guards around so they would not establish relationships
with prisoners and imposing a relentless regime of interrogation and torture.
More detainees began to confess and became informers within the camps, and
some switched sides and became interrogators or joined security force units.
While prisoners were forbidden to talk outside their own huts, they developed
a covert communication system and killed suspected spies. Living conditions
in the camps were terrible with prisoners suffering malnutrition and typhoid.
Although most detainees in the 'Pipeline' were male, including a special camp
for young boys, a few thousand women and girls were also detained.

Beginning in June 1954 and lasting for the next year and a half, the British for-
cibly resettled over one million Kikuyu into 800 'protected villages' surrounded
by barbed wire, deep trenches with spikes and watch towers, and patrolled by
the Kikuyu 'Home Guard'. Inspired by the British counter-insurgency program
in Malaya, the protected villages were divided between those suspected of supporting Mau Mau who were sometimes denied food aid and loyalists who needed
protection. In April 1954 General China surrendered and avoided execution by
trying to arrange other capitulations. When General Sir Gerald Lathbury took
command in May 1955 there were around 3000 Mau Mau in the field and the
British 39 Brigade was withdrawn by the end of the year. Lathbury phased out
large security force sweeps in favor of smaller specialized units including 'counter
gangs' or 'pseudo-teams' of former insurgents who infiltrated real rebel groups to

kill or capture them. Insurgent numbers declined to 900 in early 1956 and Dedan Kimathi, the last active commander, was captured in October, tried and hung.[3]

In 1956 the British granted a series of reforms in Kenya increasing the amount of land available to the Kikuyu in the reserves, lifting the ban on Africans growing coffee which had been the profitable preserve of white settlers, raising urban wages for Africans and permitting the direct election of African members to the Legislative Assembly. The Mau Mau uprising showed the British government that it would have to use increasingly expensive and publicly embarrassing military force to continue colonial rule in Africa. One of the most controversial incidents occurred at Hola detention camp in March 1959 when 11 prisoners were beaten to death for refusing to work and the Kenya administration unconvincingly claimed they had died from drinking poisoned water. As a result, in 1960 the British ended the emergency and announced that the concept of shared African and European 'multi-racial' rule in Kenya would be abandoned in favour of 'one-person one vote' majority rule which led to independence three years later. In conjunction with other international events such as the 1956 Suez Crisis and the rise of broad anti-colonial views, Mau Mau helped pave the way for the independence of Britain's other African territories during the 1960s. In recent years a debate has arisen over the total number of Kenyans who perished during the Mau Mau emergency ranging from estimates of 50 000 to 300 000. During the insurgency the British executed 1090 Kenyans. The insurgents killed at least 1819 African, 32 European and 26 Asian civilians. The security forces suffered 600 dead and claimed to have killed 10 500 Mau Mau in combat. The post-colonial Kenya governments of Jomo Kenyatta and Daniel Arap Moi did not celebrate Mau Mau as a national liberation movement likely because of the presence of former colonial loyalists in their administrations. More recently, Mau Mau veterans have been declared national heroes and heroines, and in 2010 the government declared 20 October, the anniversary of the declaration of the emergency, as 'Heroes Day'. In the early 2010s, the rehabilitation of Mau Mau in Kenya and the publication of several important books about British atrocities during the crisis led to successful legal action against the British government which compensated some Kenyans for suffering human rights abuses.[4]

Early Tracking Operations

During the early Twentieth Century a colonial hunting and game-keeper culture developed in Kenya that emphasized a lone and often aristocratic white man in a remote wilderness assisted by one or two loyal African gun-bearers and trackers from pastoral or hunter-gatherer communities. Beginning in the early twentieth century, Kenya was also home to a safari industry in which professional 'great white hunters' and entourages of black servants took paying foreign tourists out into the wilderness for hunting and eventually wildlife photography. These hunting ideals, much repeated and romanticized in popular literature, would have a great impact of the development of British counter-insurgency in 1950s Kenya.[5]

Early in the Mau Mau Emergency, the Kenya Police and Kenya Regiment established small and fortified forest outposts in the Kikuyu reserves and white farming areas bordering on the insurgent hideouts around the Aberdares and Mount Kenya. Manned by two white and 20 black personnel, each post mounted local patrols aimed at tracking, ambushing and killing Mau Mau fighters.[6] The basic problem in locating Mau Mau guerrillas was geographic as explained by Peter Mills, a senior officer of the Kenya Police; 'The vast areas of forest available to them made tracking and engaging them an almost impossible task'.[7] Ian Parker, a Kenya Regiment veteran, described the experience of forest patrols which characterized daily life during the conflict for many security force members:

> Forest patrolling – heavily laden with weapons and food; clambering, slithering, wet day and night for weeks on end – was an activity none would undertake for fun. Occasionally the sheer drudgery of creeping and clambering through this sad habitat looking for fresh tracks was leavened by a rush of adrenalin when really fresh sign was found, or some sound indicated Mau Mau nearby. Then the creeping and stalking had purpose, though more often than not it subsided all too soon when the tracks were lost as the Mau Mau became extraordinarily clever at concealing their trails … Months went by with no sign of Mau Mau except their tracks. Very occasionally, there was action. A flurry of shots as one or two at the head of a patrol had the briefest of glimpses as a couple of men dressed in rags and skins disappeared into the dense undergrowth incredibly quickly.[8]

When the emergency began, the British Army in Kenya quickly established formal links with the Game Department and newly created National Parks Department. At the end of 1953 East Africa Headquarters reported that 'Steps are being taken, in conjunction with the Game Department, to recruit more efficient trackers and to replace those who are found to be incompetent'.[9] Given their experience with remote areas and access to skilled trackers, many of Kenya's state game keepers were quickly enrolled in the Kenya Police Reserve (KPR) and involved in the early phase of combating the insurgency. Like other colonial civil service organizations, the Game and National Parks departments were based on a racially hierarchical structure in which armed white game wardens commanded usually unarmed black game scouts and other black employees engaged on an ad hoc basis such as trackers and skinners. In 1953 game warden Jack Sim led a unit called 'Sim Force' that consisted of ten white Kenya Regiment soldiers and several Game Department trackers which was successful in pursuing and attacking Mau Mau groups. At the start of the emergency game warden Rodney Elliott, recalled from a grouse hunt in Scotland, patrolled the Mount Kenya forests with one or two African game scouts equipped with firearms. Since the insurgents were then new to the forests, Elliott's experienced unit easily tracked them and called police and Forestry Department reinforcements to attack camps and pursue large groups.[10] In late December 1953 and January 1954 game warden George Adamson, later famous for his and his wife's work with lions as depicted in the 1966 film 'Born Free', and his African game scouts were dispatched to the Aberdares where they assisted two British battalions with tracking and directed patrols to several Mau Mau camps. Subsequently, Adamson returned to his usual jurisdiction of Isiolo in northern Kenya where, in April 1954, he was instructed to arm his 20 game scouts and help the police and army track Mau Mau insurgents who were beginning to infiltrate the area to escape security force operations to the south and acquire weapons from Somali smugglers. Called Adamson Force, the small unit was reinforced by a quick reaction team of 12 horse mounted African police from Moyale. In September 1954 the British Royal Inniskilling Fusiliers were sent to the area and Adamson Force worked as guides and trackers for their patrols. By the emergency's conclusion, Adamson Force had helped kill 16 insurgents including some shot by Adamson himself who was decorated by the governor.[11] According to a British officer, 'George excelled his own trackers when they had difficulty with the spoor and he remained unruffled, decisive and wise throughout the chase'.[12] Throughout 1952 and 1953 Fred Bartlett struggled to find time for his Game Department duties as he was increasingly deployed as a police reservist tracking Mau Mau gangs from African labour huts on European farms where they acquired supplies to their sanctuaries in the high forests. For Bartlett, evidence of Mau Mau attempts at anti-tracking often gave away their identity:

> The gangs did not like to leave their tracks on bare ground when they moved around. If there was a pathway, they stepped to the side of it. By looking carefully we could see where they had pressed the grass flat. When we saw this evidence we knew they were Mau Mau tracks and not those of law-abiding citizens. The forest areas above were closed to everybody except the security forces, so any tracks or signs of people (other than military or police) moving around were suspect. On the forest trails Mau Mau food carriers sometimes left faint signs of maize meal which trickled onto the ground as a result of holes in their bags. On two occasions I tracked food carriers in this way.[13]

At the emergency's start young Bill Woodley, assistant warden of the new Tsavo East National Park recently returned from obligatory Kenya Regiment military training in Southern Rhodesia, was serving as a platoon sergeant in 26 KAR in the Fort Hall District. After his unit encountered insurgent tracks which no one could follow, Woodley gained his company commander's permission to return to Tsavo where he recruited skilled Waata (also called Liangulu) elephant hunters Hekuta Simba and Galo-Galo Guyu who were subsequently responsible for apprehending the perpetrators of several killings around Fort Hall. Since Woodley had worked as a professional hunter in the late 1940s, he probably already knew Hekuta Simba whose father had been a famous tracker-guide for the emerging safari industry during the 1930s. While Woodley initially paid the trackers from his own pocket, these successes prompted the local district commissioner to put them on his payroll. At the start of January 1953, while working as a reconnaissance group for a company from 5 KAR in the eastern Aberdares, the Waata trackers located a Mau Mau gang which had recently killed two white farmers. Using the noise of a stream to muffle the sound of their approach, Woodley and his trackers opened fire on the group killing four and wounding one while the remaining two fled.[14] The involvement of Kenya's gamekeepers in pursuing insurgents contributed to the growth of a security force discourse that, as Wendy Webster points out, 'produced the Mau Mau as a form of savage wildlife to be tracked and killed'.[15]

Mau Mau fighters came to see African game scouts as traitors to their cause. Throughout the emergency insurgents often attacked routine African game scout patrols.[16] African game scouts working as trackers for the security forces risked retribution from Mau Mau sympathizers. Later in the war, a scout named Kamino who was based at Ragati Forest Station, respected among white hunters for his courage with dangerous animals and who had helped the security forces track insurgents, was lured to a drinking session by some forest labourers and on his way home was abducted and executed in the forest.[17]

The need to employ trackers to locate Mau Mau insurgents changed the exclusively white composition of the Kenya Regiment. This transition began with 'Intelligence Force' (or I Force) which, beginning in December 1952 at Nyeri, consisted of a small composite company of Kenya Regiment white soldiers, Kenya Police, loyalist Kikuyu, and Ndorobo, Samburu and Turkana trackers who were given some weapons training. Mounting small forest patrols each with at least two African trackers, I Force set ambushes on insurgent routes, discovered insurgent camps and focused on cutting off the insurgents hiding in the Aberdares from nearby communities that were supplying them. I Force pioneered a number of other tactics that would become standard security force practice during the emergency including Police Air Wing support of ground patrols, use of tracker dogs which in this case were on loan from the South African Police, and incorporation of Mau Mau captives who changed sides and worked as trackers. In February 1953 I Force was removed from police command and brought under its parent Kenya Regiment battalion which organized three operational companies to use the same methods. Subsequently, Kenya Regiment tracking detachments of two white sergeants and six black trackers each were seconded to every British company (there were four companies per battalion and a maximum of six British battalions in Kenya at any one time) in the country with a typical ten man British patrol accompanied by one KR sergeant and two trackers.[18] According to Parker, I Force commander Neville 'Cooper demonstrated the need for trackers and no one grasped the role good tracking played in coming to grips with Mau Mau quicker than Guy Campbell',[19] the Kenya Regiment commanding officer. Campbell maintained that 'Africans were even more at home in the terrain; trackers were found to increase efficiency by 50 per cent. By 1954 there were around 400 trackers in the Regiment'.[20] In February 1953 Game Department warden Monty Brown 'produced' the first Kenya Regiment African tracker, an African Game Department employee named Kibwezi Kilonzo from the reputedly martial Kamba ethnic group, who assisted patrols in the Aberdares.[21] Although many Kenyan whites had grown up with hunting in the bush, very few possessed refined tracking skills as they usually relied on African trackers to find game. Within African communities, people developed tracking abilities by pursuing lost livestock and/or predators, and engaging in illegal hunting. When the emergency began, some white Kenya Regiment soldiers reported for duty with their own civilian African trackers who were usually Kalenjin, Turkana or Kikuyu initially paid and supported by their private employers until enrolled as dual members of both the KPR and military. In February 1953 Woodley, recently returned to the Kenya Regiment presumably with his personal trackers Hekuta and Galo-Galo, was sent to recruit more trackers from among the Waata community which had a reputation for poaching in Tsavo National Park. Woodley first went to Malindi on the coast where he enlisted 25 Waata and Giriama hunters, and then moved to the infamous poaching centre of Mutha in Kamba territory where he recruited another 50 trackers the absence of which temporarily reduced illegal hunting in Tsavo. Other Kenya Regiment white recruits also brought in African trackers some of whom had previously served in the KAR or Kenya Police. While some Kenya Regiment African trackers were initially armed with spears, and bows and arrows, firearms training for trackers began in May 1953 and by July all were equipped with rifles. The recruitment of black trackers also enabled the Kenya Regiment to compensate for its shortage of white manpower, drawn as it was from a tiny settler minority, and deflect administrative proposals for its incorporation into the police. Among the Kenya Regiment's black trackers were some of the first Mau Mau prisoners to change sides. Officially titled 'Tracker Kenya Regiment', an African member's conditions of service were based on those of other black colonial troops such as the KAR as he could progress no higher than sergeant major and for administrative purposes belonged to racially segregated platoons though operational patrols were racially mixed. As early and particularly skilled recruits, most of the Waata became non-commissioned officers including Sergeant Kiribai Ngonyo who had fought in Burma with the KAR during the Second World War, Sergeant Galo-Galo Guyu and Company Sergeant Major Hekuta Simba. Although their service records were destroyed to avoid post-colonial retribution, it appears that at least 1500 African personal served in the Kenya Regiment during the Mau Mau Emergency which represented half the unit's operational strength.[22]

Security force recruitment of African trackers was based on colonial divide-and-rule strategy, and reflected established and overlapping stereotypes of certain African communities as naturally martial people such as the Kamba and Kalenjin who dominated local security forces like the KAR, and innately gifted trackers such as the Ndorobo and Maasai. Since these groups were also disproportionally represented in the safari industry and Game Department, the white game keepers who recruited many security force trackers tended to favour them.[23] While the martial groups within the Kenya Regiment amounted to 34 per cent Kalenjin, 18 per cent Kamba and 10 per cent Turkana and Samburu, some 20 per cent were drawn from the reputedly non-martial Kikuyu, Embu and Meru who were associated with Mau Mau.[24] An Oxford anthropologist celebrated that 'The old enemies of the Kikuyu are now being trained as Police Askaris, and their remarkable tracking abilities are a great asset in a war where men must be hunted in the thick forests that clothe the Aberdares. The Masai, the Kipsigi, the Turkana and the Samburu put it this way: "Thank God. The white men have finally come to their senses and are helping us kill the Kikuyu."'[25] However, much of the campaign took place in high altitude forests where local knowledge and acclimatization became important factors. For example, the Kenya Regiment's 'I Force', in March 1953, imported 120 Maasai hunters from the grasslands to sweep the mountainous forest of the Aberdares but they could not tolerate cold weather and were quickly sent home.[26] Major Gordon L. Potts, a veteran of the Korean War and an officer with 7 KAR in Kenya from 1954 to 1956, observed that Maasai trackers were excellent in the open plains but useless in the forests where operations took place.[27] Kenya Regiment commanding officer Campbell wrote 'we had recruited trackers from the other tribes used to forest conditions. To start with we had Masai – they were lifelong enemies of the Kikuyu – but they were not in their true element in the forest. The best forest and bush fighters proved to be the El Geyo, Turkhana, Nderebo, Wakamba, Nandi, Samburu and loyal Kikuyu.'[28] Although they lacked a martial or tracking reputation, Kikuyu men from the Kikuyu Home Guard or former Mau Mau became important as trackers. A report on Kikuyu loyalists stated that 'This was the country in which the Kikuyu Guard excelled. They knew it intimately. They could track; they could think like the enemy; they could, lightly equipped, move fast by night or day'.[29]

Not all African trackers who worked for the security forces were motivated by financial gain. Loyalties between white hunters and black trackers from marginalized minorities remained strong. At the start of the rebellion Gichimu, an Ndorobo honey-collector who had worked at the same white owned farm at Kinangop in the Aberdares since 1906, refused to take the Mau Mau oath and volunteered as a tracker for the army and police. He declined pay or other compensation and continued to track for the security forces until the end of the emergency when he was awarded the British Empire Medal. While the sixty-something year old Gichimu tracked unarmed and often alone, and believed that the Mau Mau were misguided youth who should be captured and convinced to change their ways, he led security force patrols to rebel hideouts where insurgents were killed or wounded. Gichimu often tracked with Kenya Police Reservist Venn Fey who had grown up on the farm where he worked and who he had taught to track as a child in the surrounding forests. Fey later wrote that Gichimu was 'as courageous as a buffalo, loyal to his friends ... Possibly one of the greatest and most widely known of all African trackers during the Emergency'.[30]

Tracker Training

In November 1953 East Africa Command opened a Tracking School at Nanyuki, an established military centre in a white farming area northwest of Mount Kenya, to train soldiers and police arriving from Britain. The Kenya Regiment provided the initial Tracking School staff which designed its training curriculum within the context of local conditions. Five out of the six founding instructors worked for the Game Department or National Parks in civilian life. The Game Department employees included Rodney Elliott, commissioned into the Kenya Regiment given his experience as a British commando during the Second World War, who became the school's commander and instructors Monty Brown who had learned tracking as a boy from a Kikuyu Ndorobo in the Mount Kenya forests, George Adamson and Don Bousfield. The single National Parks instructor was Peter Jenkins from Tsavo who, like Woodley, had recently returned from military training in Southern Rhodesia. Furthermore among the instructors was Jim Tooley, a legendary tracker in his mid-50s who had been raised by the Kipsigi in the Chepalunga Forest and worked as a farm manager. Tooley recruited four Kipsigi trackers to assist with tracking instruction. His knowledge of several African languages including Kikuyu enabled him to recruit African trackers for British units, and eventually tour prison camps to select and train former Mau Mau as trackers. Although the illiterate Tooley remained an 'acting sergeant', he became highly sought after as a tracking advisor by British units in the field and spent much of his spare time working voluntarily in that capacity. While the first tracking course in Nanyuki began with a few lectures on tracking basics, handling African trackers, local wildlife and use of tracking dogs, most of the time was spent on practical demonstrations and at times police reports enabled the students to track real insurgents which resulted in some skirmishes.[31] The use of hunters and gamekeepers in such training was not entirely novel as during the Second World War's Burma campaign the British Army appointed 73 year old honorary Lieutenant Colonel Jim Corbett, a tiger hunter in northern India and author of several popular hunting books, to teach soldiers about jungle survival and tracking.[32] Kenya Regiment patrol leader Lieutenant Lenard Gill, an early student of the Tracking School who later became an instructor there, was not completely impressed by the training:

> I expected to see some pretty hot demonstrations from the instructors, but I was disappointed. Except for one (probably Jim Tooley), the instructors were babes in arms when

compared with Ngalu my brilliant Mkamba tracker, under whose tutelage I had become proficient. The main advantage I gained from attending the course was points, where my comprehension had been vague, were clarified. The flow of knowledge from other patrol commanders was perhaps as useful as the training received from the instructors, who were more practised in following animal tracks than those made by humans.[33]

In late 1954 the Tracking School was absorbed into a new and larger East African Battle School also based at Nanyuki. Since tracking became part of a wider training program that also included bush warfare and use of tracking and patrol dogs, and British regular army instructors arrived, most of the Kenya Regiment personnel were deployed elsewhere.[34] It should be pointed out that while tracker dogs were widely used by British forces in Kenya, some officers considered them a nuisance as they became exhausted too easily.[35] The Battle School created its training program by assembling the seven most successful patrol commanders in Kenya who developed standard procedures and then became instructors in anti-Mau Mau operations. The 'Magnificent Seven' believed the best way to eliminate an insurgent group in the forest was for one tracking team to follow it and use a radio to inform another team of its intended route upon which an ambush would be prepared perhaps by a small unit transported by vehicle along one of the cleared forest tracks.[36] British personnel and former Mau Mau insurgents working for the security forces learned tracking together and practised by trying to follow trails set by school personnel. The tracking section of the Battle School assessed African trackers to determine if they possessed suitable skills and taught them how to apply such knowledge to counter-insurgency work, and it prepared European junior leaders to supervise trackers. According to chief tracking instructor Gill:

> Our aim was to make students capable of managing trackers. They had to be able to determine the skill of a tracker, whether he was doing his job or was deliberately 'losing' the spoor. Not all former Mau-Mau men were keen to operate against their one-time comrades or lead a patrol when there was a danger of being ambushed. The Battle School lectures helped patrol commanders and trackers by teaching the theory. Skill in the art of tracking would burgeon with experience.[37]

7 KAR officer Potts found the tracking course useful as it taught him broader lessons such as that soldiers hunting insurgents in the bush had to abandon customs like regular tea breaks and live like their prey.[38] The Battle School instructors helped write anti-Mau Mau training manuals which included sections on tracking, Mau Mau anti-tracking methods, Mau Mau symbols left in the bush as a form of communication with other groups, potential encounters with dangerous animals, and basic Kiswahili vocabulary and phrases. The influence of hunters/game keepers is obvious in the introductory remarks from an anti-Mau Mau manual:

> The Mau Mau are fleet of foot, silent in movement, highly experienced in fieldcraft and normally anxious to avoid action with all forms of organized Military Forces.

The qualities which must be developed in troops engaged against the Mau Mau are therefore those required to track down and shoot sky game.[39]

At the start of 1954 every British battalion in Kenya was meant to have at least 30 local African trackers and this was later expanded to 36. At first, the British officers of the KAR battalions did not want local trackers as they believed their African soldiers naturally possessed these skills. However, this changed when KAR officers observed trackers at work during Operation Anvil and from that point each KAR battalion was assigned 15–20 of them. The number of local trackers was lower than for British battalions 'Since most KAR battalions have a few askari who are competent trackers'.[40] J.J. Hespeler-Boultbee, a platoon commander with a Tanganyika battalion of the KAR (probably 26 KAR), recounts that:

> In the months patrolling the forests of the Aberdare mountains I became passably good at tracking. Several of my askaris, Nyamahanga (Warrant Officer Platoon Commander) among them, were highly skilled at it, and I was naturally curious to be able to see what they saw. There were so many signs to look for – a scuff mark, a turned leaf or blade of grass could open up a whole new avenue; the sounds of animals moving in the near distance, or the sudden flight of a bird, could be an indicator they were being disturbed by others if not by the passing of one of our patrols. Body odour or farts could hang a surprisingly long time in hot, motionless and dusty airs.[41]

Peter Brind, commanding officer of 5 KAR in Kenya from 1954 to 1956, pointed out that his African soldiers were 'automatic as trackers' and their ability to determine if a mark on a forest trail or tree had been made by an animal or person represented an important advantage over British soldiers. He also suspected that Scottish soldiers who had grown up in the highlands had similar abilities.[42] In May 1955 KAR personnel who worked as trackers and passed tracking tests at the Battle School were given additional daily pay based on their skill level. 70 Brigade, made up entirely of KAR battalions, arranged for 168 KAR soldiers to take these tests at a rate of 30 per day and the Battle School also launched a series of week-long special tracking courses for KAR soldiers.[43]

Pseudo-Teams

One of the most well-known British tactics developed during the Mau Mau Emergency involved pseudo-terrorist teams consisting of disguised security personnel and eventually captured insurgents who had changed sides who attempted to infiltrate insurgent groups with a view to eliminate them. This tactic began with security force night patrols making simple attempts to impersonate Mau Mau to test the loyalty of Kikuyu communities. During the last half of 1954 there were two simultaneous experiments with 'pseudo-gangs'. Kenya Regiment intelligence officers Francis Erskine and later Stan Bleazard, and British District Military Intelligence Officer Frank Kitson began cultivating teams including captured Mau Mau to engage in elaborate insurgent impersonations meant to locate and penetrate enemy groups. While Erskine and Kitson cooperated and exchanged converted insurgents, the increasingly successful pseudo-operations were eventually centralized under a new Special Force in 1955. Kitson later wrote a book on this tactic which was used in subsequent counter-insurgency campaigns such as in Malaya, Northern Ireland and Rhodesia.[44] Within the Kenya Regiment, National Parks warden Woodley played a pioneering role in developing the pseudo-gang concept which he credits to fellow soldier Steve Bothma. According to Woodley, the first use of this tactic in Kenya occurred in October 1954 in the Kiambu area when a team consisting of himself, Bothma and a Kikuyu tracker disguised themselves as insurgents to approach and then open fire on a Mau Mau group. The whites blackened their faces and walked behind the black tracker who took the lead in making contact with the enemy.[45] Daphne Sheldrick, Woodley's wife at the time, explained that African trackers were central to the early pseudo-teams:

> Bill and Francis (Erskine) operated with a team of specially chosen men, some drawn from the Waliangulu (Waata) elephant-poaching fraternity, whom Bill had recruited as trackers from areas bordering Tsavo. Bill had great respect for these expert bushmen, as proficient at tracking humans through the forest as they were at following a dikdik under desert conditions. Mau Mau converts who had turned informer were also a crucial part of Bill's team. They knew the forest intimately – every path, every glad, every ravine, the location of the hideouts, the clandestine ceremonial meeting places, the hollow forest trees in which communications passed between the rebel commanders. Above all, they knew the secret signals and calls used by the Mau Mau; a stick tossed on a path at a certain angle; the leaf of a particular plant left on the ground; a twisted stem, a pebble or two – all conveyed a specific message, as did various sounds, which no outsider could decipher as anything other than the call of a bird or animal. It was these Mau Mau defectors who were best qualified to bring about the downfall of the hardcore within the forests.[46]

Woodley's National Parks experience was relevant to these under-cover operations since just before the emergency he and some of his African assistants had impersonated illegal ivory dealers to collect intelligence on poachers operating in Tsavo East. Moreover, state game keepers like Adamson and Woodley had often convinced poachers they had arrested to turn their knowledge and hunting skills against their former colleagues by becoming game scouts much as captured Mau Mau were convinced to change sides.[47] Tracking represented one of the main methods used by the pseudo-gangs to find their targets. The Kenya Police also formed pseudo-teams which, according to Mills, 'would track gangs through the forests and when they eventually came into contact, the terrorists were fooled into thinking that they'd met up with another group of Mau Mau and were immediately mown down. Mau Mau groups were made so nervous by these activities that they often ended up shooting at each other'.[48]

Tracker Combat Teams

In early 1955, after several large scale yet unsuccessful operations in the Aberd-ares and Mount Kenya forests and a failed Special Branch attempt to negotiate a mass insurgent surrender, the British approach to the war shifted to using small Tracker Combat Teams (TCTs) and pseudo-gangs to isolate and eliminate the remaining Mau Mau.[49] While the development of pseudo-teams has received considerable attention from historians, the formation and use of TCTs are not well understood. Planning for the transition to small units began many months before. Indeed, at the beginning of December 1953 East Africa headquarters informed brigade commanders that patrols and ambushes were not having much success as they were often detected and avoided by insurgents. A study conducted in Malaya reached similar conclusions. Most of the blame was put on British soldiers' poor noise discipline and bush marksmanship, and lack of skilled trackers. The three brigade commanders in Kenya were told that 'It is apparent that the only real way in which the percentage of success in patrols and ambushes can be materially improved is by beating gangs at their own game'. They were instructed to explore and report back in six weeks on the feasibility of forming a 15 man 'commando-type' unit in each brigade. These units were to develop a high level of bush craft such as silent movement, remain in the field for long periods with light loads, report information and location quickly presum-ably by radio and possibly 'include a white tracker'.[50] These concerns informed the creation of the East Africa Battle School which was meant to cultivate bet-ter bush warfare skills among the security forces. It appears that Rodney Elliott, founding commander of the Tracking School, first proposed the concept of a Tracker Combat Team (TCT) consisting of a tracking section of three African trackers under a European leader, a support group of four or more soldiers to do the fighting, and a dog section with a patrol dog for early warning of ambush and a scent tracker dog. Elliott also insisted that teams become intimately familiar with a single area by working there for a long time and that six teams operate together under a Tracker Group commander which would allow them to coor-dinate their efforts by, for example, concentrating to assault a large Mau Mau camp or using fresh teams to spell off exhausted ones during a pursuit or set ambushes along likely trails.[51] While brigade and battalion commanders initially wanted to keep the TCTs separate from each other and under control of their parent battalions, Elliott objected strongly and was supported by East Africa headquarters which steered the brigades towards centralizing the teams.[52] In July 1954 each of the 13 major units in Kenya, 11 infantry battalions plus an artillery battery and armoured car squadron, were instructed to form a TCT. Selected by East Africa headquarters from among suitable volunteers, the team command-ers were sent to the units to personally select their team members and initiate

several weeks of preparatory 'hardening' and bush shooting practice which was
then taken over by the team's second-in-command. The Battle School's Tracking
Wing dispatched three African trackers to each new TCT. As the teams were
being prepared, the team leaders undertook a three week tracking course at the
Battle School after which they returned to their teams and put them through a
final three weeks training in the operational area where they would be working.[53]

In mid-August 1954 East Africa headquarters informed the commander of
49 Brigade that 'we must polish up our jungle craft to compete with the increasing cunning exercised by the Mau Mau in covering up their tracks in the jungle'.
49 Brigade was to concentrate its five recently deployed TCTs, hitherto spread
out with three around Fort Jericho and two around Kiambu which was a populated area in which tracking was pointless, and place them under Venn Fey who
'has done extremely good work in the Western Aberdares and is quite obviously
a highly skilled tracker of the Rodney Elliott calibre'.[54] Fey was a white Kenyan
who owned a farm at South Kinangop in the Aberdares, a veteran of the Second World War's East African campaign, and spoke Kikuyu and Kiswahili. As a
Kenya Police Reservist, he had spent the early days of the emergency interrogat-
ing prisoners until a nervous breakdown prompted him to spend several months
recuperating in Scotland. One of the 'few White men who were outstanding
trackers', he had influenced the creation of TCTs by reporting that 'most of the
patrolling that is being done is of little value ... This, beyond doubt, is principally
due to the fact that the European officers leading patrols have not the faintest
idea what to look for, or how to differentiate between an ordinary game track and
a Mau Mau track'.[55] Fey was commissioned as a captain in the Kenya Regiment
specifically to lead 49 Brigade's experimental Combat Tracker Group and on the
condition that he could resign whenever he wanted. Fey also brought along his
childhood tracking mentor; the Ndorobo honey-collector Gichimu.[56] During
September and October Fey led a group of three TCTs into the south Aberdares.
On Elliott's advice, this rainy season was the best time to track insurgents as they
left behind more obvious signs in the wet ground and forest though these would
be wiped away immediately after a rainfall. Fey's group first concentrated on the
bamboo forest area at the headwaters of the Thika River which had, since the
beginning of the emergency, been considered a centre of Mau Mau organization
in the Fort Hall area. Working in concert, Fey's teams found evidence of extensive
Mau Mau occupation including many camps and a hospital but it appeared all
had been abandoned about three weeks earlier given aerial bombardment. Since
the effectiveness of bombing had been questioned, this information was surpris-
ing and useful. By mid-October Fey's TCTs had killed, wounded or captured 27
insurgents. The next major target of Fey's group, now reduced to two TCTs plus
three additional Kikuyu Home Guard trackers, was the forest bordering a series
of African reserves around Fort Hall based on intelligence that insurgent groups

had moved there from deeper in the forest. Patrolling across the grain of the country to locate tracks, Fey's men spent two weeks pursuing several small Mau Mau groups and a larger one of 100 insurgents which they contacted on several occasions but always escaped. At the end of October Fey led both his teams, one from the Royal Northumberland Fusiliers and another from 6 KAR, in an assault on an insurgent camp that killed six insurgents and wounded another 16 who fled into the forest.[57] The British press quickly reported that of 27 insurgents killed by security forces within a 24 hour period, the new TCTs had accounted for 16.[58]

In mid-November 1954, given Fey's success, East African headquarters instructed all three brigades in Kenya to group their TCTs with a view to dominating a specific area of reserve and adjacent forest that they would each be assigned after the upcoming Operation Hammer. With a view to expanding the program, it was expected that each battalion including those rotating into the country would create two TCTs by the start of January 1955. KAR battalions were given some flexibility on this deadline as they lacked enough suitable European leadership to assign two Europeans to every team which Elliott considered essential, especially for the KAR. Headquarters pointed out that 'The Commander-in-Chief is convinced that the best and most economical way of dealing with prohibited areas is by using tracker/combat teams on long duration patrols'.[59] During November and early December 1954 staff officers at headquarters began planning the formation of commando-style companies each consisting of three or four combat tracker platoons (essentially TCTs) and a 'Trojan' or pseudo-terrorist platoon. The commando company would still belong to a battalion for administrative purposes but its company commander would take orders directly from his brigade headquarters. Trojan teams would operate more independently in conjunction with military intelligence and police Special Branch. Fey and Kitson, the respective TCT and pseudo-team experts, were brought to a conference in December to discuss these plans. Efforts by battalion commanders, eager to retain their authority, to form one tracker platoon per Rifle Company were resisted by headquarters as it was thought much depended on the leadership of a company commander, modelled on Fey, directing several small tracker platoons.[60] In practise, this resulted in each battalion forming a Forest Operating Company with a number of TCTs though the Trojan teams appear to have been left out of the new structure. Forest Operating Companies would patrol primarily in the forest or forest edge where tracking could work effectively as there were few inhabitants while the Trojan teams would work in the populated reserves. In January 1955, during Operation Hammer which involved a massive security force sweep of the edge of the Aberdares forest, Fey commanded 49 Brigade's prototype Tracker Combat Group which operated 3000 yards ahead of the main force and killed 12 Mau Mau, five of which were personally dispatched by Fey on a single day. Since Operation Hammer employed nine infantry battalions and resulted in the death of 161 insurgents at a cost of over £10 000 each, the work of Fey's comparatively tiny group in accounting for almost ten percent of enemy losses confirmed its effectiveness and cost efficiency. Subsequently, Fey was promoted to major to reflect the new Forest Operating Company structure and he eventually received the military cross.[61] During February 1955 Fey personally trained the new Forest Operating Company commanders in the Aberdares while their team leaders undertook a course on trackers and dogs, and team sergeants who led assault sections took another course on forest battle drills and quick forest shooting at the Battle School. These personnel then returned to their Forest Operating Companies which, during March, undertook patrol training in the Aberdares. By 1 April there were 20 trained TCTs in Kenya grouped into four or five companies. This training cycle was repeated to create more teams with Fey continuing to mentor new company commanders and prospective team leaders temporarily attached to existing teams to gain experience.[62] Spending three weeks in the field and one in camp or at his farm per month, Fey later wrote about this period that 'I enjoyed tracking. The job was pure hunting, the enjoyment not being in the kill, but in pitting one's skill against a cunning and elusive quarry'.[63] Around this time, Fey broke military discipline by publically criticizing the administration's policy of granting amnesty to former Mau Mau who had abandoned the struggle.[64]

During most of 1955 there were ten Forest Operating Companies in Kenya; five from British battalions, four from KAR battalions and one from the Kenya Armoured Car squadron. Each company consisted of at least three TCTs with each team comprising a junior officer or NCO who was in charge, a radio operator, eight soldiers, two or three African trackers, a tracker dog and a patrol dog and, if needed, an interpreter.[65] The TCTs were employed for 'special tasks requiring a high degree of fieldcraft and endurance', and served as a model for all the parent unit's other patrols to emulate.[66] To some extent, each parent battalion interpreted the Forest Operating Company/TCT concept in its own way. During March and April 1955 the Royal Green Jackets transformed its 'B' Company into a Forest Operating Company with each of the three platoons divided into a 8–10 man 'patrol' and an 8–10 man TCT, the only difference being that the latter possessed a tracking dog. In preparation, four of the company's junior officers and sergeants had undertaken an exhausting three week combat tracking course under Fey. The company worked out of a base camp at Squire's Farm in the Aberdares foothills with personnel not assigned to a patrol or TCT formed into a 'passive wing' (a term usually associated with Mau Mau) for logistical support and local security. According to one of the company's officers, 'it was essential for our patrols to be able to live in the forest for long periods at a time without Mau Mau being aware of their presence. Gangs had to be hunted down by methodical patrolling across the grain of the country to find trails used by terrorists. The trails were then ambushed or more usually followed up, the African trackers attached to each patrol searching each exit from the track for signs of a terrorist hide'.[67] Given that many of its personnel were supporting tracking training and operations across all elements of the security forces, the Kenya Regiment was reduced to a single operating company with a special 'Tracker-Combat' force of two teams each with a white sergeant and private, three black trackers, and a tracker dog with an army handler that was on stand-by for rapid deployment.[68] The Kikuyu reserves, by June 1955, had been largely pacified by the security forces and the Forest Operating Companies aimed to track the now smaller Mau Mau groups and drive them out of the forest where they could be more easily located and eliminated.[69]

The pseudo-teams became part of a different organizational structure. In May 1955 General Lathbury received permission from the War Council to establish five Special Force Teams (SFTs) to operate in conjunction with the military and police Special Branch. Each SFT was commanded by a European and consisted of ten former Mau Mau insurgents who, as special constables with one month's training, would track and pose as insurgents. Such small formations of indigenous trackers were considered essential in hunting the remnants of Mau Mau.[70] However, these SFTs still had trouble tracking the insurgents. In late October 1955 team leader Captain R.J. Folliott, after conducting a reconnaissance of the Thiba River area, concluded that 'Owing to heavy rains and game, tracking was practically impossible.' In late January 1956 Folliott, hunting for Dedan Kimathi in the northeast Aberdares, reported that 'Despite intensive attempts to find tracks this was impossible. It is obvious that the enemy are becoming increasingly efficient in concealing their movements'.[71]

By June 1955, given that civilian African trackers were increasingly difficult to find and pseudo operations had illustrated the usefulness of former insurgents, the Battle School had begun to recruit Mau Mau prisoners as trackers primarily by having Tooley visit detention camps and select those who seemed to have potential. The likely candidates were taken to the Battle School for testing where, for example, 19 out of 50 in June and 20 out of 23 in July were sent to units for employment as trackers. None of the former insurgents received the top of three tracker ratings and Tooley believed that Special Branch was withholding the best material for the pseudo-gangs. Former Mau Mau trackers tended to be deployed to units working in their home areas which they knew well. Around this time the 205 African trackers attached to army battalions represented a mix of civilians and ex Mau Mau with, for example, 1 Gloucestershire Regiment (Glosters) having 36 trackers of whom 11 were converted insurgents. The Royal Irish Fusiliers and Glosters considered that most of their trackers were 'NBG' or 'No Bloody Good', and 4 KAR complained that its civilian trackers were 'not first class' and requested 20 former Mau Mau. A proposal from someone at headquarters to recruit trackers from Bechuanaland (Botswana), home of the renowned Bushmen trackers, was dismissed as 'not worth the expense' as the emergency was winding down.[72]

Police Tracker Teams

The formation and development of Kenya Police tracker teams was directly related to insurgent stock theft from white farmers. In April 1954 a delegation from the Stock Owners Association of Kenya visited the acting governor to point out that since June of the previous year some 1621 cattle and 987 sheep had been stolen with the government paying £60 000 in compensation. It was noted that the vast majority of thefts had happened in the Mweiga/Ngobit area. Colonel Melvin Cowie, head of the National Parks Department, became involved and convinced five white professional hunters to take their own African trackers to that area during April and May, an off season for the safari industry, where they recovered 336 out of 473 cattle reported stolen at that time. Given support from white farmers and the fact that most of the hunters were over the age of compulsory military service, the administration, in June 1954, approved the permanent deployment of five tracker teams in Mweiga/Ngobit. Personnel for the teams were sought from among white professional hunters, the Game Department, National Parks, white farmers over military age and the security forces, particularly the Kenya Regiment. Although most likely candidates were already engaged in emergency duties including as Tracking Wing instructors, nine white hunters volunteered for the teams by August and were enlisted as police reservists. In September the Minister of Defence arranged for Captain Jack Bonham, an experienced hunter recently retired from the Game Department, to visit the area and he recommended the establishment of eight tracker teams each with two Europeans, two African trackers and four African policemen supported by a vehicle and radio. This was approved by the War Council and by the middle of November five such tracker teams were operational. By the middle of December all eight teams were active with five based at Nanyuki and three at Nyeri. Since they represented an important part of Kenya's economy, the white professional hunters leading these tracking teams could take leave when they were commissioned to lead a tourist safari. Many of the white hunters in these teams had limited direct experience with tracking and relied on African trackers. An exception was Peter Becker who was such a good tracker that he became overworked during the emergency. Disapproving of the role of civilian white hunters, the Kenya Police sent contract inspectors fresh from Britain and other members of the police reserve on tracking courses at the Battle School and attached them to existing teams to gain experience before taking over. The police planned to expand the number of tracker teams to 19 by disbanding its mounted unit and converting two existing 'hunting patrols' in Laikipia and Naivasha. Furthermore, police authorities were concerned that these small tracker teams did not have sufficient firepower to survive an encounter with a large Mau Mau gang.[73] It appears that these police tracker teams eventually borrowed the name TCT from the army though they were not formally grouped together like Forest Operating Companies. They pursued a 'dual role of recovering stolen stock and destroying the gangs responsible for the thefts'.[74] Police veteran Mills wrote, 'our forces became expert reading the signs and silently tracking enemy gangs'.[75]

Police TCTs often cooperated with military units and exploited recently captured or surrendered insurgents to locate and attack Mau Mau groups. One of the team leaders found that pursing Mau Mau stock thieves usually resulted in one of two situations. Firstly, they would find a large pile of offal where the animals had been butchered which meant it would be easy to track insurgents whose feet left deeper impressions in the ground as they were encumbered with loads of meat. Secondly, a small group of Mau Mau would often slaughter stolen animals and hide the meat somewhere nearby taking great care to conceal signs of this activity and then purposely leave obvious signs of movement away from the scene including dropping pieces of meat on the trail. In this case, the police tracker team would depart to convince any hidden Mau Mau sentries that they were continuing the chase but surreptitiously leave behind several members to ambush anyone returning to recover the meat.[76] Derek Franklin, a Kenya Policeman who took over Becker's accomplished tracker team in 1956, observed that the basic security force tactic at the beginning of the insurgency involved large numbers of personnel sweeping the forest in the hope of driving Mau Mau fighters towards other police and soldiers deployed in stop lines. Franklin maintained that 'the concept of 'Combat Tracker Teams' came much later; but when it came it proved to be far superior to the old ways and was a natural progression as past experiences were analysed and digested'. However, Franklin's assessment of the performance of his own tracker combat team in countering Mau Mau stock theft from white farms on the lower slopes of the Aberdares around Mweiga and Ngobit was less positive: 'Our efforts to combat the threat were largely ineffectual. It was more a case of following up after an incident, sometimes following tracks for half a day or more, but never achieving a satisfactory contact'.[77] In 1954 and 1955, game wardens Bartlett and Bousfield, after the latter had recovered from being shot during a tracking operation, formed a full-time police tracker combat team that operated on the north side of Mount Kenya. Bartlett remembered that 'The various District Commissioners sent us numerous candidates from the Masai, Lumbwa and Ndorobo tribes. Eventually, we selected six likely candidates, hoping to mould them into successful trackers'. Bartlett and Bousfield armed their trackers with rifles from the local police armoury and taught them to shoot. 'Afterwards we took them out tracking and found a couple who were really good at it. Then came the real thing. We tracked gangs several times and actually made contact a few times'.[78] Tim Symons, a young British settler who joined the Kenya Police and took over a tracker team when the white hunters went back to their safaris, remembers that it was a 'romantic and prestigious' position that gave him 'film star status' at bars and hotels in Nairobi. He also recalls that his skilled Maasai trackers were specifically recruited because they hated the Kikuyu and were considered higher status than an ordinary African soldier or policeman and had more money than most people in their community.[79]

Insurgent Anti-tracking

Mau Mau insurgents were keenly aware that the security forces were tracking them. According to Mau Mau leader General China, 'We learned, too, to walk through the forest with great care, leaving no traces of footprints or broken twigs...' and that discarding 'a spent match could put the enemy on our track'[80] When anti-tracking, Mau Mau insurgents walked backwards to make it appear as if they were moving in the opposite direction, took high and long steps to minimize the number of tracks and had the last man in a group brush away footprints with a branch. They also walked on stilts to leave no footprints, used vaulting poles to jump sections of open ground or obstacles, stepped on blankets which were then removed, and famously disguised their footprints by wearing elephant or rhino feet.[81] Referring to this last method, KAR Warrant Officer Nyamahanga maintained that 'this is always obvious ... Those who have passed are easily given away by the depth of the indentations in the earth. An animal places its hoof squarely on the ground; a human has to roll his foot, back-to-front'.[82] Some anti-tracking was more lethal as Mau Mau fighters sometimes created a false and obvious trail into the forest edge and then circled back to set an ambush for security force trackers. Odor was also important as the insurgents smelled like the bush in which they lived while police and soldiers often smelled like soap, hair-products or tobacco.[83] Varying by area, insurgents created a code of symbols they would leave behind in the forest to tell allies about direction or give warning. These included bent or broken twigs or leaves, marks on bark, purposely positioned quills and holes dug along a path which usually indicated the presence of a hidden food cache. In addition, as they became aware of the existence of pseudo-teams, they employed whistles and animal calls to identify each other in the dark.[84] Among their white adversaries, Mau Mau gained a reputation for highly refined bush skills which was related to the idea that they had reverted to an animal-like state. Police tracker team leader Hewitt maintained that 'Mau-Mau bushcraft and ability to interpret the minutest disturbance of grass or soil was absolutely phenomenal. In an instant a 'wary' mick (Mau Mau was derisively called Mickey Mouse) could decide whether the being that made the mark or disturbance was friend or foe. There was a method, or better still, a code, of movement and passage through the forest when in the vicinity of the hide that was practised by Mau Mau and which defied imitation'.[85]

Subsequent Counter-Insurgency and Anti-Poaching Operations

The experience of recruiting, training and employing trackers during the Mau Mau emergency influenced British counter-insurgency elsewhere and there were proposals to dispatch skilled Kenyan trackers to other theatres. During Mau Mau a British officer from the operational research unit in Malaya visited Kenya to study local techniques and he was particularly impressed by the work of African trackers. As a result, 'on his return, the whole question of tracking in the Malayan jungle was studied in great detail' and a dedicated 'Tracking Wing' was created at the British Army Jungle Warfare School where Iban trackers from Borneo – similarly stereotyped as natural warriors like the Kamba and Turkana in Kenya – were introduced to local conditions and patrol commanders learned how to best utilize joint tracking by humans and dogs. An army scientist told a reporter that 'Unless a commander understands the elements of tracking himself and can "talk turkey" to his tracker, he will not achieve very much'.[86] In January 1958 the First New Zealand Regiment in Malaya formed a battalion tracking team that was always on standby to respond to the scene of an insurgent contact and separate company tracking teams for less urgent tasks. Similar to the tracker combat teams pioneered in Kenya, these teams consisted of Iban trackers, surrendered enemy personnel, Malayan interpreters and New Zealand dog handlers with tracking and early warning dogs. The battalion team was commanded by Huia Woods, a former school teacher and veteran of earlier service in Malaya with the New Zealand SAS, who had become a legendary tracker and reputedly the only person to attain an 'A' grade for tracking at the Jungle Warfare School. While the Iban became central to these operations, Woods considered some of the Malayans and New Zealanders to be better trackers. Further experience with counter-insurgency tracking in Borneo and Vietnam in the 1960s prompted the New Zealand SAS to retain tracker training years after these wars had concluded.[87] Although some explain the New Zealand military focus on tracking as originating from the Maori heritage of many of its members, this is questionable as the Maori did not have a strong tradition of hunting and tracking given the country's lack of large mammals. Partly Maori himself, Woods also rejected this claim pointing out to fellow soldiers in Malaya that he had grown up in the city and explaining his affinity for tracking as based on logic taught by his white parent.[88]

In 1956 an officer from East Africa Command visited Cyprus and advised officials there to employ trackers in hunting down insurgents in the mountains. Since there were no local trackers, Cyprus Governor John Harding requested that the Kenya administration lend him some white trackers from the Kenya Regiment as black ones would be unacceptable to the racist population of Cyprus. He also asked for a dozen Turkish Cypriot police to undertake tracker training in Kenya. Although Kenya Governor Barring was sympathetic to the proposal, most of the Kenya Regiment and KPR members had gone back to their civilian lives and white professional hunters would expect a very high fee to travel to Cyprus. As such, the only Kenyan trackers to serve in Cyprus were dogs. However, some Cypriot police did go to Kenya for a seven week tracking course.[89] A similar proposal to dispatch Kenyan trackers to Oman proved more successful. In late1958 Colonel David Smiley, the British commander of the Omani Armed Forces who owned a farm in Kenya and was married to a white Kenyan, directed intelligence officer and Mau Mau Emergency veteran Frank Kitson to go to Kenya to recruit trackers from the Northern Frontier Province which was geographically similar to arid Oman. Smiley and Kitson had failed to find any suitable Omani trackers. On the advice of Kenya's Chief Game Warden, Kitson visited wardens Adamson at Isiolo and Elliott at Maralal who recruited ten local trackers. Under the supervision of Stan Bleazard who had served in the Kenya Regiment during the emergency and was now given leave from the Kenya Post Office, six Kenyan trackers were flown to Oman where they were employed in trying to trace the planters of landmines back to their villages and in training Omani trackers. In 1960 increased minelaying incidents prompted British authorities in Oman to request and receive a second contingent of Kenyan trackers.[90] This influenced Kitson to suggest that the British army form a small special unit of trackers recruited from different environments around the world that could be deployed with the first troops to arrive in an operational theatre until the services of reliable local trackers could be secured. It appears that this plan was never put into action.[91]

Security force tracking operations against Mau Mau greatly influenced Kenya's post-colonial anti-poaching campaign. While Kenya's game-keepers contributed their bush skills and contacts with expert African trackers to the anti-Mau Mau campaign, they learned military and counter-insurgency skills and tactics that were subsequently applied against poachers. As explained by historian Ed Steinhart, the Mau Mau war 'proved a school and a training ground for the key personnel and may well have helped inspire the strategies, and prepare the ground for the next phase of anti-poaching operations'.[92] In 1955, as the emergency was winding down, Kenya's National Parks Department formed the Voi Field Force under David Sheldrick and Bill Woodley to mount a sustained campaign against the poachers around Tsavo. Many of the early Kenya Regiment African trackers during the emergency who had been recruited by Woodley joined the National Parks service as members of the Voi Field Force including Hekuta Simba who became a senior sergeant and Kiribai Ngonyo.[93] Instrumental in launching the program was Noel Simon who, after serving as a fighter pilot during the Second World War, became a farmer and wildlife enthusiast in Kenya and during the emergency he served in 'I Force' and then managed the KPR Air Wing which conducted aerial reconnaissance. He founded the Kenya Wildlife Society in 1955 and joined the upper ranks of the National Parks Department the following year.[94] The military culture of this anti-poaching force resembled the KAR with Woodley commanding a platoon of 30 African soldiers from imagined martial peoples who were divided into three ethnically oriented sections; Somali, Samburu and a mixed one of Nandi, Turkana and Kamba. An intelligence section, working directly under Woodley and consisting of some of the former Kenya Regiment trackers and several turned Waata poachers, began to build detailed files on individual suspects. In October 1956 the successful anti-poaching campaign under Sheldrick was expanded to include six European officers and 100 African troops organized into three field forces based at Voi, Hola and Makindu, supported by a police spotting aircraft and prosecutions officer. Like many captured Mau Mau who had changed sides, apprehended Waata and Kamba hunters often confessed to poaching and were then convinced to join the National Parks anti-poaching force where they used their tracking skills against their former associates.[95] The program's effectiveness was facilitated by the expansion of National Parks authority outside park boundaries, an increase in prison terms for those caught with poaching paraphernalia such as arrows and poison, and the lack of legal restrictions such as search warrants. All the Europeans involved in leading the late 1950s anti-poaching campaign had experience fighting Mau Mau including David McCabe, Ian Parker and Dennis Kearney from the Game Department, Peter Jenkins from National Parks and Major Hugh Massey from the Kenya Police who had commanded Sheldrick's mounted unit during the rebellion.[96] McCabe's emergency experience was particularly impressive and relevant as he had pioneered pseudo-teams as a Kenya Regiment field intelligence officer in Embu District in 1954 and the next year was appointed commander of one of the new Special Force teams.[97] According to Steinhart, what started as the Voi Field Force 'would become the model for the post-colonial anti-poaching units created with World Bank assistance and operated in Kenya from the late 1960s to the present'.[98] After Kenya's independence in 1963, white veterans of the Mau Mau emergency continued to shape anti-poaching operations there and in neighbouring countries. Jack Barrah began the emergency as a Kenya Regiment sergeant attached to the Lancashire Fusiliers as a tracker, guide and interpreter, and ended as a district officer leading Maasai warriors against insurgents. He joined the Game Department in 1956 as George Adamson's apprentice and eventually became post-colonial Kenya's chief game warden in the early 1970s.[99] Between 1960 and 1970 Bleazard, after his stint in Oman, worked for the Kenya Game Department where his pilot training was supported by funds from the Adamsons' lion project. Subsequently, he worked briefly as a pilot and administrator under Sheldrick at Voi in Tsavo National Park, and during the 1970s he held senior positions in the national parks departments of Zambia and Malawi.[100] The military role of game-keepers did not end with independence as Elliott and the Game Department assisted Kenyan security forces during the counter-insurgency campaign, also known as the Shifta War, along the Kenya-Somalia border from 1963 to 1967.[101]

Conclusion

While the most well-known experimental small unit to play a role in the demise of Mau Mau was the pseudo-team which is popularly associated with Frank Kitson, the tracker combat team conceived by Rodney Elliott and pioneered by Venn

Fey also made an important contribution. The mountain forests of the Aberdares and Mount Kenya were difficult places to track Mau Mau insurgents who developed a high level of anti-tracking ability but the limited, though still large, size of the operational areas and declining number of insurgents, given the collapse of their civilian support, facilitated the process. In Kenya, the main method for security tracking teams to catch up with insurgents was to employ several teams together with an exhausted pursuit team being replaced by a fresh one and radio communication used to send other teams ahead, often by motor vehicles driving on tracks cut in the forest, to intercept the prey. Aerial reconnaissance by small planes was also helpful. Recruiting of trackers was done from communities with a reputation for this skill such as the Kamba, Kipsigi and Ndorobo whose homes tended to be located outside the operational area but more and more local Kikuyu were employed in that part given their local knowledge and familiarity with Mau Mau. In the tracking teams, which were influenced by local colonial hunting culture and the Nanyuki Tracking Wing, Europeans were leaders and supervisors while African trackers did most of the work of following signs. Though few in number, state game keepers were essential in leading, training and recruiting trackers, and developing tracking tactics. Unlike future counter-insurgency campaigns in late twentieth century Africa to be discussed below, scent tracking and patrol dogs were widely used in Kenya and this was likely because the cold temperatures of the high forests allowed scent to settle on the ground whereas in hotter areas it quickly evaporates. The British Army also had established expertise and resources related to military and police dogs. Given the success of British counter-insurgency against Mau Mau, aspects of the campaign would be copied in other conflicts including the emphasis on tracking.

4 RHODESIA (ZIMBABWE), 1965–80

During the 1890s the British South Africa Company (BSAC), controlled by Cape-based mining magnate and ardent British imperialist Cecil Rhodes, occupied the area between the Limpopo and Zambezi rivers inhabited primarily by the Shona and Ndebele peoples. BSAC forces invaded and conquered the Ndebele Kingdom in 1893 and subdued a rebellion by elements of the Ndebele and Shona in 1896–97. The territory became known as Southern Rhodesia while the area colonized by the BSAC north of the Zambezi became Northern Rhodesia. Since the vast gold deposits anticipated by Rhodes turned out to be highly exaggerated, the Southern Rhodesia's economy became dominated by large white settler commercial farms with Africans mostly pushed into small reserves, later renamed Tribal Trust Lands (TTL), from where they provided cheap labour. When the BSAC withdrew from administration in 1922, the territory's small white population rejected a proposal to join the Union of South Africa and accepted responsible government, a form of internal self-government, from Britain which was granted the following year. Although Southern Rhodesia adopted a non-racial but qualified voting system, racial discrimination and the nature of the colonial economy meant that very few blacks qualified to vote. The white minority used its control of the state to pass laws favourable to itself such as the Land Apportionment Act of 1931 which restricted African land ownership to a very small portion of the country. A white Rhodesian identity arose and many settlers looked forward to further political devolution from Britain along the lines of the autonomous dominions such as South Africa and Canada, and it was believed that Southern Rhodesia had earned this anticipated status by the participation of its soldiers in both world wars. At the same time, and mirroring developments across the continent, an African national-ist movement emerged that began in the early twentieth century with small groups of westernized elites who peacefully petitioned the state to reduce racial discrimi-nation and turned more radical and expanded across urban areas immediately after the Second World War. With the election of the anti-British Afrikaner National Party in South Africa in 1948 which implemented a strict form of racial segregation called apartheid, London sought to create a new powerful ally in Southern Africa by combining the small territories of Southern Rhodesia, Northern Rhodesia and

Nyasaland into the Central African Federation in 1953. Britain invested heavily in the federation helping to build a massive hydro-electric project at Kariba on the Zambezi River and expanding federal military air power. As the federation was controlled by the tiny white minority of Southern Rhodesia, African nationalist movements in all three territories staged increasingly violent protests which were crushed by security forces in a series of states of emergency at the end of the 1950s and beginning of the 1960s. This internal turmoil, coupled with the move towards decolonization across Africa, prompted Britain to disband the federation in 1963 and the next year independence and majority rule political systems were granted to Northern Rhodesia which became Zambia and Nyasaland which became Malawi. In those countries, the African nationalist movements that had protested colonial rule and federation entered government and quickly turned into autocratic one party states. Given that Southern Rhodesia still enjoyed responsible government, Britain could not legally interfere in its internal affairs and the local white minority elected the conservative Rhodesian Front to maintain their dominance. This led to a political impasse in which Southern Rhodesia demanded the immediate and unconditional granting of dominion status which Britain refused unless there was a guarantee of eventual universal suffrage. In November 1965 the Rhodesian Front government, led by Ian Smith, enacted a Unilateral Declaration of Independence (UDI) from Britain which did very little to stop this illegal action. Smith's Rhodesia became an international pariah and the target of gradually increasing economic and military sanctions. However, it survived by building close ties with apartheid South Africa and colonial Portugal with its territories in neighbouring Mozambique and Angola, and using its strongly anti-communist stance to appeal for limited covert support from the Western world including the United States.

The dissolution of federation precipitated the division of the Zimbabwe African People's Union (ZAPU) the leadership of which had gone into exile in newly independent Zambia to avoid arrest at home. Some of nationalist activists remained loyal to the group's original leader Joshua Nkomo who wanted to remain in exile and mobilize international sanctions against Rhodesia. However, others wanted to return home to lead the people in protest against the settler regime and formed a new organization called the Zimbabwe African Nationalist Union (ZANU) under Ndabaningi Sithole and later Robert Mugabe. Many of the top leaders of both groups, including Nkomo, Sithole and Mugabe, returned to Southern Rhodesia in 1964 only to be imprisoned for the next decade. The Rhodesian break with Britain prompted the divided Zimbabwe nationalist movement, with many of its top leader detained, to form armed wings and launch separate armed struggles from across the northern border in Zambia where they received Eastern Bloc support. ZAPU founded the Zimbabwe People's Revolutionary Army (ZIPRA) and ZANU established the Zimbabwe African Nationalist Liberation Army (ZANLA). Since both groups lacked sufficient military preparation or resources for protracted war, the initial campaigns of the late 1960s were poorly planned and aimed at frightening the Smith regime into negotiation and satisfying external sponsors like the Organization of African Unity (OAU) which consisted of independent African states. During the first phase of the war in the late 1960s ZAPU infiltration from Zambia was done in conjunction with South Africa's exiled African National Congress (ANC) which sought to move insurgents through Rhodesia and enter apartheid South Africa which was surrounded by a buffer of friendly or dependent states.

At the time of UDI, Rhodesia had a small and relatively new but effective security force. For most of its history, the territory's primary defence force had been the British South Africa Police (BSAP) which was also a law enforcement organization. The BSAP was white led, consisted primarily of unarmed black regular constables and could fall back on a white reserve. It also maintained a small paramilitary unit of armed black police to be called on during emergencies. In 1947 the BSAP was withdrawn from defense duties to concentrate on policing, and a permanent military establishment was created with the only major full-time unit being an infantry battalion of the Rhodesian African Rifles (RAR), demobilized immediately after the Second World War but then quickly reformed, with white officers and black rank-in-file. During the federal period white officers from Southern Rhodesia took command of a larger army which consisted primarily of three British colonial African infantry battalions in Northern Rhodesia and Nyasaland, and the RAR. In the early 1960s intensifying African nationalist protest and an army mutiny in newly independent Congo prompted white federal officials to establish several new and exclusively white military units recruited from Southern Rhodesia, South Africa and Britain. Around the same time, the BSAP was generally expanded and countered nationalist protest by creating an African police reserve in black urban areas. With the break-up of the federation, military assets were divided among the three territories but Southern Rhodesia retained the lion's share including most of the new white units and newly acquired aircraft. In 1965 the Rhodesian army consisted of a full-time component of two infantry battalions, the all-white Rhodesian Light Infantry (RLI) and predominantly black RAR, and an all-white Special Forces unit called the Rhodesian Special Air Service (SAS). In addition, young white men were obliged to undergo a period of military training and subsequent annual call-up within several part-time infantry battalions of the Rhodesia Regiment. One of the most potent in the region, the Rhodesia Air Force flew up-to-date jet fighters and bombers, light helicopters and transport aircraft. Although the old BSAP para-military element was greatly expanded to several battalions over the next decade, Rhodesian officials were hesitant to do the same with the RAR given their long-held fear of arming black soldiers.

African nationalist insurgents seeking to infiltrate Rhodesia to conduct guerrilla warfare faced some daunting challenges. Geographically, the centre of

Rhodesia was dominated by a high plateau which forms the watershed of the Zambezi River in the north and the Limpopo in the south. Given its reduced prevalence of tropical disease, the plateau became the location of most of the territory population centres and white owned commercial farms which were attractive targets for armed nationalists. Formed by the Limpopo River, the southern border with powerful apartheid South Africa was entirely out of the question for insurgent activity. In the north, the imposing physical obstacle of the Zambezi River including Victoria Falls and the massive man-made Lake Kariba demarcated the entire border with Zambia, and much of the flat floor of the Zambezi Valley and the slopes of the Zambezi Escarpment consisted of a series of large unpopulated game reserves and national parks such as Chizarira and Mana Pools. Although Zambia was the only adjacent state to Rhodesia that openly supported the nationalist movements in the 1960s, its common border was certainly the most difficult to infiltrate. Western Rhodesia, which was close to the great Kalahari Desert and bordered on Botswana, was characterized by arid and sparse woodlands, and much of the northwest had become the enormous depopulated wilderness of Wankie (now Hwange) National Park in the late 1920s. For insurgents, this was an inhospitable area to cross and the cautious African government of Botswana, which was the British colony of Bechuanaland up to 1966, was hesitant to allow them to use its territory as it feared reprisals from militarily stronger Rhodesia which also controlled the country's railway. The best place for insurgents to enter Rhodesia would have been the eastern border with Mozambique as it was dominated by a series of mountain ranges containing grasslands, tropical rainforest in some valleys and higher dry forests. Furthermore, both sides of this frontier were inhabited by a substantial Shona-speaking rural population among whom insurgents could conceal themselves and gain support. However, at the time the Rhodesian insurgency began in the late 1960s Mozambique was ruled by colonial Portugal which was fighting a counter-insurgency war against local African nationalist guerrillas and, as we will see, forged a strong military alliance with Ian Smith's government.

This situation in Mozambique eventually changed. Beginning in 1964, FRELIMO used staging areas in southern Tanzania to infiltrate northern Mozambique where it launched a guerrilla campaign and within a few years had expelled Portuguese forces from some areas and created liberated zones. In 1968 FRELIMO opened a second front by moving its fighters through Zambia to cross into central Mozambique's Tete Province and threaten strategically important Zambezi River crossings and the Cabora Bassa hydro-electric project. Although a Portuguese offensive mounted in 1970 called Operation Gordion Knot aimed to destroy FRELIMO bases in the north and cut off infiltration routes from Tanzania, it ground down because of politically problematic Portuguese casualties and heavy rains that hampered logistics. The success of a FRELIMO offensive in Tete during the early 1970s led to Portuguese reprisals including the massacre of several hundred civilians. In 1972 elements of ZANLA, under the military leadership of Josiah Tongogara, moved from Zambia to Tete in Mozambique where it allied with FRELIMO and began to learn from this more experienced movement. That year ZANLA re-launched the insurgency in Rhodesia by infiltrating across the forested and hilly northeast border with Tete and attacking a white farm. Following the 1975 assassination of ZANU leader Herbert Chitepo in Zambia which was blamed on internal disputes but was likely carried out by Rhodesian agents, ZANLA moved entirely to Mozambique. When the beleaguered Portuguese suddenly withdrew from Africa and FRELIMO took over Mozambique in 1975, the conflict in Rhodesia intensified and expanded as ZANLA began using the entire eastern border as a staging area.

From around the early 1970s, the two Zimbabwe liberation movements developed along different lines. Remaining in Zambia, ZAPU-ZIPRA gained Soviet support and adopted a Leninist strategy to develop a conventional mechanized brigade that could invade Rhodesia at a critical moment though the geography of the Zambezi River border rendered this strategy questionable. Those ZIPRA fighters who infiltrated Rhodesia focused on building support through ZAPU's existing political networks and directly engaging security forces and establishing liberated zones. As it was based in Zambia to the north and, from 1974, Botswana in the west, ZAPU-ZIPRA recruited mostly from Rhodesia's southwestern Matabeleland region which meant that its composition became dominated by the local Ndebele and Kalanga people. With Chinese sponsorship, ZANU-ZANLA adopted Maoist revolutionary strategy that emphasized politicizing the rural masses and fighting a guerrilla conflict that would progressively lead to a takeover of centers of power. Throughout the 1970s the Mozambique-based ZANLA worried less about fighting Rhodesian security forces and more about using persuasion and coercion to gain backing from rural people. Among the most popular methods of political indoctrination was an all-night meeting called 'pungwe' in which insurgents would lead local people in revolutionary songs and statements, and denunciation of traitors who would often be tortured and killed. Throughout the 1970s villagers were expected to provide food, information and recruits to ZANLA insurgents or face reprisals. Additionally, ZANLA tried to identify with rural people by emphasizing grievances over land and employing traditional Shona spirit mediums, and the ethnic Shona character of the eastern region came to dominate the movement. ZANLA developed a highly organized structure and employed very small insurgent cells that, for security reasons, did not know about one another.

Losing control of the countryside, Rhodesian security forces in the 1970s employed counter-insurgency methods inspired by 1950s Malaya and Kenya, and adjusted for local conditions. To prevent them from supplying insurgents, rural people were herded into foul 'protected villages' where they were subject to a curfew and abuse from guards. Large tracts of land were declared 'free fire zones' meaning anyone caught inside was considered an insurgent and killed. During the early 1970s the Selous Scouts was formed under Lieutenant Colonel Ron Reid Daly to employ 'turned terrorists' (captured insurgents who changed sides) to infiltrate insurgent cells and call in air mobile 'fire forces' to eliminate them. A typical 'fire force' consisted of a helicopter borne commander circling the suspected enemy position, several four man teams landed by helicopter to block likely insurgent escape routes and a larger unit of perhaps platoon strength parachuted from an airplane and then sweeping toward the target. At first the 'fire forces' where manned by the exclusively white RLI but as the war spread the white officered but primarily black RAR was given parachute training and assigned this duty. At the end of the 1970s the South African military also formed 'Fire Forces' employed in Rhodesia. Furthermore, Rhodesian security forces tried to kill insurgents by circulating poisoned clothing and tinned food as well as transistor radios containing bombs in rural areas. Rhodesian counterinsurgency stressed racking up an insurgent body count and this was accelerated in dramatic but internationally condemned cross border raids on guerrilla bases in neighboring countries such as at Nyadzonya and Chimoio in Mozambique in 1976 and 1977, respectively. However, many rural non-combatants were killed by mistake and this pushed the increasingly brutalized rural people toward the insurgent cause. Although the core of the professional Rhodesian security forces were African soldiers and police, the Rhodesian Front government hesitated to expand its armed African personnel until the late 1970s by which time it was too late. The first black army officers were commissioned in the late 1970s when the war was almost over. Chronically short of manpower, Rhodesian forces relied on young white national servicemen (conscripts) and white reservists who were called up for increasingly long periods.

After the Portuguese departure from Africa, both sides in the Rhodesian war came under pressure to negotiate. The Rhodesian economy was being damaged by international sanctions and increasing 'call ups' of whites. South Africa, upon which Rhodesia was entirely dependent for key imports such as oil and military support, experienced a surge of internal African protest from 1976 and saw the Rhodesian conflict as a threat to its new policy of dominating black ruled neighbors through economic strength. The OAU and the Zambian government were concerned that the war was crippling the Zambian economy that depended upon copper exports through Rhodesia and South Africa. In 1974 failed talks led to the release of the top nationalist leaders from Rhodesian prisons including Nkomo of ZAPU and Mugabe who took over ZANU from Sithole who had reputedly denounced the armed struggle. During the 1978 'internal settlement', South Africa and Smith arranged for the installation of a new pro-western moderate black government in which whites retained considerable power but this proved unacceptable to the liberation movements and the war continued. At the British sponsored Lancaster House talks in 1979 agreement was reached in which Zimbabwe would formally gain independence in 1980 with universal adult suffrage. Concessions to the white minority included no forced land redistribution for the first decade of independence. It was at this point that ZANU became known as ZANU-Patriotic Front (PF) after a failed attempt to unify with ZAPU. Under Commonwealth supervision, insurgents reported to assembly areas for planned absorption into the new Zimbabwe security forces which also included former Rhodesian personnel. The 1980 election was won by ZANU-PF under Mugabe who formed a government and became Zimbabwe's first prime minister.[1]

Early Counter-Insurgency Tracking

Two important aspects of white Rhodesian identity influenced the development of Rhodesian counter-insurgency tracking during the late 1960s and 1970s. White Rhodesians, most of whom had migrated to the territory in the 1950s, generally held martial and frontiersman ideals that venerated the memory of colonial conquest of the 1890s and was believed to differentiate the colony from other parts of the British Empire. If the imagined aristocratic white settler minority of Kenya represented the officers' mess of empire then white Rhodesia was the sergeants' mess which reflected its 'hands on', rough and ready attitude. The heroes of early white Rhodesian history included R.S.S. Baden-Powell, Frederick Russell Burnham, Frederick Selous and others who were seen as expert trackers, hunters and bush-fighters, and they served as strong manly role models for white Rhodesian society. As already mentioned, white Rhodesia long distrusted the idea of arming and training black soldiers which dated back to the mutiny of black police during the 1896–97 Ndebele and Shona rebellion. During both world wars, the Rhodesian administration initially portrayed the conflicts as 'white man's wars' in which blacks would produce resources for the war effort but as the conflicts dragged on the continued participation of the colony depended on the formation of black infantry battalions that were sent to distant battlefields though immediately demobilized upon return home.[2]

Some Rhodesian soldiers had developed tracking expertise during the British counter-insurgency campaign in 1950s Malaya. Eager to participate in wider British imperial operations, Rhodesia had sent two units to Malaya; an all-white squadron that joined the British SAS and the primarily black RAR battalion. Since several of the white Rhodesians who had fought in Malaya went on to top positions in the Rhodesian military including army commanders Peter Walls and John Hickman, the experience of that campaign would greatly influence Rhodesian counter-insurgency during the 1970s. Describing the situation in 1950s Malaya, Rhodesian SAS historian Barbara Cole explains that 'At first, the

Rhodesians were confused when it came to tracking. It was completely different from tracking back home where the sun was a vital factor and where the dust, sand, rock, dry leaves, wild animals and their habits all helped. In the jungle, the packed foliage blotted out the all-important sun and sky, reducing daylight to twilight...'[3] While the Rhodesian SAS troopers attended the British Jungle Warfare School and were accompanied on patrol by Iban trackers brought in from Borneo, the unit established its own tracking training area in Dusan Tua where types of footprints were displayed and, more importantly, the men learned how to estimate the age of a trail cut through thick foliage by studying examples purposefully cut at different times. Three man patrols were assigned specific areas and upon returning to camp submitted a 'Track Report' which indicated what signs of insurgent movement had been observed. Locating insurgents in the dense jungles of Malaya proved extremely difficult as a patrol that was only 30 minutes behind an insurgent group would struggle to catch up. One of the Rhodesian SAS men later wrote that 'deep in the jungle any footprint had to be regarded with suspicion in what were in fact "no-go" areas'.[4] Later in the 1950s RAR African soldiers with previous hunting experience at home were given specialized tracker training in Northern Rhodesia (now Zambia), then part of the Federation of the Rhodesias and Nyasaland, and at the Jungle Warfare School in Malaya and embarked on operations as 'scout-trackers'. Obert Veremu, a RAR veteran of Malaya, explained that 'to be a scout you must be very clever, it is difficult, you follow spoor and find scratches in trees. I was very keen to get these bandits'. Veremu recalls that a typical jungle patrol lasted from 14 to 20 days and that on one occasion he spent four days tracking a single enemy group but lost them as 'the insurgents were very clever'.[5] Paul Mufanebadza, an RAR soldier who had grown up hunting and fought in Burma during the Second World War, was employed as a tracker in Malaya and says that his unit's expertise in this field was used to produce a training manual for British forces.[6] According to W. A. Godwin, an RAR officer in Malaya, African soldiers:

> were good in the bush or jungle for they would see things to which we Europeans were simply not attuned. Most of them were reasonably good trackers and some were brilliant. This gift they no doubt learned when as herd boys they were required to track down and find lost cattle from the family herd.[7]

For John Essex-Clark, an RAR platoon commander in Malaya and later an Australian officer in Vietnam, 'A well-trained African platoon moved very fluidly and fast in the jungle compared with British platoons but not as fast as the terrorists. The askari (African soldiers) could track well, and we were often forewarned by keenly observant scouts who had fought in Malaya before'.[8] Lieutenant D.K. Bales, a white officer with the primarily black Northern Rhodesia Regiment (NRR) in Malaya, claimed that 'The African soldier is psychologically suited to the jungle, which although novel holds few terrors for him, which arouses his hunting, tracking and practical instincts'.[9] The Malaya experience taught the Rhodesian Army the importance of tracking in counter-insurgency and training pamphlets were produced that explained pursuit operations but afterwards 'nothing was done to teach tracking'.[10]

The pioneer of military oriented tracking in the Central African Federation was Allan Savory who, in the mid-1950s, was a Northern Rhodesian Game Department official and Territorial Army soldier. Much later, an American protégé wrote:

> Allan obsessed over the bush in his boyhood Rhodesia. He taught himself to track, pushing himself to the point of coming to the end of a lion trail – that is, flushing the lion – after hired-gun professionals had quit the trail because of the danger of darkness and thickening vegetation. Allan tracked the rogue cattle-killer on hands and knees, "feel tracking" in the dark under the leaves, not to destroy the lion but to prove that a teenager who could master fear could "out-track" fearful professionals.[11]

While tracking poachers and problem lions as a game ranger, Savory noticed that African trackers under his supervision often became fearful and lost the spoor when they closed in on a dangerous opponent. 'After a few such episodes, I realized I simply had to become a better tracker myself ... I spent endless hours simply determined to be the best I could in all aspects of bushcraft and certainly to match any native tracker'. During the 1959 Nyasaland Emergency, which was meant to suppress African nationalist protest in that territory, Savory noticed that his fellow all-white territorial soldiers from Southern Rhodesia possessed very few bush skills. Seeing that the federation might soon face armed insurrection from increasingly violent African nationalist groups, Savory read voraciously about guerrilla warfare and questioned his colleagues who had fought in jungle campaigns against the Japanese during the Second World War. In 1960 Savory, who had transferred to the Southern Rhodesian Game Department and become a Territorial Army officer, ambitiously wrote Federal Prime Minister Roy Welensky recommending the military prepare for impending insurgency by training military trackers but this was rejected by conservative senior officers. Savory's ideas slowly gained acceptance when, in the early 1960s, the new and exclusively white Federal Army SAS invited him to give lectures and short courses on bush survival including tracking. During a field exercise near Nkomo Barracks in Southern Rhodesia in the early 1960s he personally demonstrated combat tracking by apprehending four mock guerrillas while a white territorial infantry battalion located none. This impressed Malaya veteran Essex-Clark, now a regular force training officer with the territorials, and Savory gained official permission to conduct longer courses for the SAS in the Zambezi Valley on what was now called 'aggressive bushcraft' in which tracking was applied to warfare. With the end of the federation in 1963 and Southern Rhodesia's UDI in 1965, it was clear armed insurgency was inevitable though many of the SAS officers trained by Savory left the country rather than serve the rebel Smith regime.

In 1965 Savory, now a private game rancher in Southern Rhodesia but still a territorial officer, presented a paper to Rhodesian Army Headquarters suggesting the formation of a Guerrilla Anti-terrorist Unit (GATU) with white SAS operators and black police who would specialize in tracking and anti-tracking, and pose as insurgents to infiltrate and eliminate their groups. Although Savory received permission and conducted 'top secret' selection courses in an unpopulated part of the Sabi Valley, disagreement between the army and police led to the sudden demise of the embryonic unit. Subsequently, in the late 1960s the army allowed Savory to select white territorial soldiers who were professional hunters and game rangers in civilian life to form the Tracker Combat Unit (TCU) – sometimes called Combat Tracker Unit or CTU). Lacking black personnel, the planned infiltration of insurgent groups was set aside. Most TCU members were retrained in tracking as in their civilian careers they had relied on African trackers which meant their skills had not developed. Utilizing four man teams, the unit developed standard procedures for tactical tracking and emphasized silence, instinctive shooting, long distance tracking and fast movement on foot to catch fleeing prey. To enhance silence so as not to alert an enemy, TCU members wore shorts and t-shirts so they would avoid rustling through thorn bushes, carried their rations in bandoliers crossing the chest instead of regular army web equipment that could get caught on bush, and communicated by hand signals and special low-tone dog whistles. Indeed, reminiscent of persistent hunting by early humans, it has been suggested that the TCU's skimpy attire was also meant to enhance the body's natural cooling system. Anti-tracking techniques were taught and TCU members wore special canvas hockey boots with custom designed soles inspired by the feet of an elephant which leaves little sign on the ground despite its weight. While Savory knew that scent tracking dogs had been used by British counter-insurgency forces in Kenya during the Mau Mau Emergency, his experience with them in the Game Department and professional hunting informed his decision not to employ them within the TCU. Dogs made noise that would warn an enemy of the presence of trackers, they obliterated spoor, and they required food and water. In theory the full-time SAS focused on external operations while the part-time TCU worked inside Rhodesia responding to insurgent sightings and trying to track them. Given Savory's efforts, around 1966 the SAS became the first Rhodesian regular army unit to establish formal tracker combat teams the nucleus of which consisted of personnel who had been professional hunters in civilian life. These included Darrell Watt who, at 17 years old, had been hired to cull wildlife on a private farm and as an SAS operator became one of the first army trackers to track insurgents to the point where they were engaged. Over the next few years the SAS became the 'tracking pool' for the Rhodesian army which somewhat distracted it from the main focus on Special Forces-style clandestine missions.[12] For Paul French, a member of the Rhodesian SAS in the middle 1970s, the unit's emphasis on spending long periods in the bush searching for insurgent spoor was 'boring, tiring and rarely yielded any concrete results'. French sought to transfer out of the SAS because he believed it was not engaged in true Special Forces-style missions but functioned more like a 'Parachute Regiment Patrol Company'.[13]

As stated previously, the location of ZAPU and ZANU staging areas in Zambia during the late 1960s presented problems for infiltrating neighboring Rhodesia. The two countries were divided by the physical obstacles of the Zambezi River and Lake Kariba, and Rhodesia's northern Zambezi Valley was far from likely insurgent targets and its sparse and un-politicized African population could not be relied on for support. Furthermore, the area's large and uninhabited parks were patrolled by rangers; water sources were limited in the dry winter and thin tree cover made aerial observation a possibility. In the remote Zambezi Valley, according to Rhodesian military historian J.R.D. Wood, 'fresh human tracks command instant attention'.[14] To make matters worse, ZAPU/ANC insurgents who crossed into Rhodesia in 1967 and 1968 traveled in relatively large groups of up to 100 and all wore the same Cuban-made boots that made a distinctive 'figure 8' pattern footprint which facilitated security force tracking. According to Thula Bopela and Daluxola Luthuli, veterans of the disastrous ZAPU/ANC Luthuli Detachment of 1967, 'We didn't know much about tracking and back-tracking in those days and took no precautions'.[15]

The first instance of combat tracking in the Rhodesian War took place on 11 August 1966 when a group of six ZAPU insurgents who had crossed Lake Kariba heading for Matabeleland were followed and captured by two civilian geologists and a detachment of five black RAR soldiers.[16] Exactly a year later, the first TCU operation happened when a white game ranger in the Zambezi Valley reported seeing the footprints of around 100 guerrillas who turned out to be the ZAPU/ANC Luthuli Detachment. The TCU subsequently located and scouted several insurgent camps which were then destroyed by elements of the RAR and RLI in what became called the 'Wankie Campaign'. During the operation TCU member Joe Conway pursued four fleeing guerrillas for four days over 60 kilometers until they surrendered. Their lack of readiness for bush warfare was illustrated by a prisoner's complaint about being tracked like a wild animal. In December 1969 the TCU, responding to an insurgent attack on the Victoria Falls airport and railway, struggled with heavy rains washing away tracks but eventually discovered a camp which had been hastily abandoned by 22 guerrillas who split up and fled, and were eliminated within the next few days. Involved in almost every incident of insurgent infiltration over the next few years, the TCU lost only one man and was responsible for dozens of insurgent deaths and captures.[17] However, the austere survivalist mentality instilled in the TCU by Savory, which required soldiers to live off the land even when food was available and rejected blankets during cold nights, and constant military call ups caused resentment and some men left the unit.[18]

The BSAP, Rhodesia's law enforcement organization, also applied tracking to counter-insurgency. The organizational culture of the BSAP was slightly different than the Rhodesian military. It was a racially hierarchical organization in which whites were senior to blacks but the latter had long represented the majority of the force. In 1964 BSAP member Bill Bailey, a veteran of the British Long Range Desert Group (LRDG) during the Second World War, formed a Tracker Combat Team with volunteer police and police reservists in the Lomagundi District in the northern part of the country. This local part-time tracking unit was quickly disbanded on orders from disapproving BSAP authorities but continued informally under the auspices of Volunteer Advanced Training. With the dissolution of Savory's combined army and police GATU, Bailey's experiment influenced the 1966 creation of the country-wide Police Anti-Terrorist Unit (PATU) similarly consisting of white and black police and police reservists who, in addition to their regular duties and civilian jobs, volunteered for rural patrols to gather intelligence and pursue insurgents. PATU's volunteer and localized nature meant that its training and employment of trackers was ad hoc. Some PATU members already possessed tracking skills gained as civilian hunters or game rangers, Africans with tracking skills including Bushmen from the southwest were encouraged to join by enrolling in the African police reserve and African civilian game-trackers were sometimes employed.[19] Among the white farmers in PATU was Gysberd Smit of the small community of Banket who, in the late 1960s, used his 'exceptional tracking skill and patience' to locate an insurgent group and then informed the security forces.[20] Another PATU member was Italian professional hunter Giorgio Grasselli who owned a game farm near Wankie National Park, and during his brief initial training, 'On tracking in the bush, his personal métier, he was able to show his instructors a thing or two'.[21] BSAP officer David Lemon described a PATU unit of white farmers as 'all experienced hunters and born to the veldt'.[22] Another BSAP element to which tracking became important was Ground Coverage which, beginning in the 1960s, dispatched small teams of white and black personnel to collect intelligence from specific rural areas over a long period. Usually, their only training consisted of the standard ten day BSAP counter-insurgency course and tracking expertise was often acquired on the job or by engaging civilian specialists.[23] By 1974 the BSAP basic counter-insurgency course included a section on 'animal tracks and different types of human ones' and this was re-emphasized in more advanced patrol training for the para-military Support Unit in which white leaders considered the supposed 'well-developed visual acuity' of the mostly African members to represent an advantage in tracking insurgents.[24]

The BSAP also assembled an ad hoc force of civilian pilots and aircraft, called the Police Reserve Air Wing (PRAW), which became skilled at tracking insurgents from the air.[25] Since the BSAP had begun using scent tracking dogs to apprehend criminals in 1948, security force insurgent tracking operations in 1967 and 1968 included police tracker dogs but the results were mixed. In an encounter with the Luthuli Detachment a police dog handler was killed and his dog fled. Although the BSAP continued to use tracker dogs throughout the war, they were never widely adopted by Rhodesian counter-insurgency forces as they became exhausted by heat and rough terrain, and scent evaporated in the mid-day heat. Attempting to accelerate tracking and avoid ambushes, the Rhodesian Air Force and BSAP, during 1968 and 1969, experimented with equipping a trained tracker dog with a harness radio used to give basic orders and an orange panel to allow it to be followed by a helicopter carrying its handler and troops to be deployed on the ground when insurgents were found. The project failed as regular police dogs were expected to fill the role of tracker dogs.[26]

In the late 1960s the Rhodesian army depended heavily on civilian trackers including Bushmen from the southwest and government personnel from the Department of National Parks and Wildlife Management (DNPWLM). Indeed, the insurgent infiltration routes from Zambia during the early part of the war meant that they often had to cross vast national parks which put them on a collision course with DNPWLM personnel assigned to protect these areas. The LuthuliDepartment was first detected by National Parks ranger Dave Scammel and his black game scouts who had noticed a change in the behavior of elephant and zebra which the insurgents had begun to hunt for food which led to the discovery of their characteristic footprints and a discarded Russian sugar packet. Like other elements of the Rhodesian state, the DNPWLM was a racially hierarchical organization in which white Wildlife Officers (WOs) with ranks such as ranger or senior ranger were usually in charge of a station with a staff of black game scouts. Since the WOs were white, they were subject to periods of obligatory military service including call ups and some were active as volunteer territorial soldiers including many in Savory's TCU. When security force units operated within National Parks they were assigned WOs as trackers who were almost always accompanied by unarmed black game scouts who received extra money. In this situation WOs were sometimes working as DNPWLM employees and sometimes as territorial soldiers. During the early phase of the conflict the DNPWLM, in addition to the army's TCU, formed its own counter-insurgency Volunteer Tracking Unit (VTU) or Parks Tracking Unit that dispatched teams of one WO and one or two game scouts to aid the security forces. As an operational team, the unarmed black game scouts conducted the actual tracking protected by the armed white WO. The VTU began in the late 1960s with Paul Coetsee, provincial warden for Mashonaland North based in Sinoia and a founding TCU member, lending

National Parks trackers to the police to help apprehend criminals and insurgents. This arrangement was formalized in the early 1970s with the spread of the insurgency to the country's northeast and the launching of Operation Hurricane in that area. National Parks VTU teams took turns on two to three week operational assignments in 'hot areas' and by 1976 there were always at least three such teams at work in different parts of the country. When the TCU was disbanded in 1974 (see below), some of its former members who were also National Parks staff continued as part-time military trackers in the VTU.[27] According to former WO Kevin Thomas, 'The game scouts used were always those who knew the white ranger, had hunted regularly with him and trusted him implicitly'.[28] Around 1972 National Parks WOs, particularly Mike Bromwich and Robin Hughes who were also members of the army TCU, were involved in pioneering pseudo-teams which patrolled the operational area disguised as insurgents to collect information and catch the enemy off guard.[29] During the 1970s, with the abandonment or militarization of some parks, the WOs at Mana Pools were assigned six white national service soldiers on a six month rotation to help with security and train as trackers. With the intensification of the war in the late 1970s white National Parks officials stationed at vulnerable parks on the borders like Chizarira and Mana Pools were exempt from obligatory military service and VTU assignments as their regular patrols supplied information on signs of insurgent activity. In the mid-1970s, as wartime needs gradually overshadowed white fears of expanding black military recruitment, African game scouts tracking for the security forces were armed and given camouflage uniforms.[30] During 1978 Ranger Ron Selley provided basic tracking instruction to Rhodesian soldiers based in the southeastern Gonarezhou National Park and accompanied National Parks and security force reconnaissance flights that monitored the fenced and mined border with Mozambique which included culling elephant herds threatening to damage the barrier.[31] National Parks trackers also worked with the police. For example, they conducted tracking courses for PATU personnel in places such as Mana Pools National Park.[32]

Remembering his involvement in 1968's Operation Cauldron, former RLI lance corporal Denis Croukamp wrote that:

> For me the excitement of the chase kept me on a permanent high ... I was also putting into practise a fairly new skill I had been trained to do. I had recently just completed a second tracking course at Kariba and was gaining further experience by watching and talking to the old tracker who was assigned to us. Whenever I was on point I would walk next to him and ask him questions. Eventually he would point out and explain things without me asking him.[33]

Although Croukamp obviously learned from the black civilian tracker, his account of Operation Cauldron also illustrates the problems of employing non-military personnel in this capacity:

The tracker, a fairly old man, really started to get excited and displayed a lot of nervousness, stooping and trying to look ahead, at the ground and at me all at the same time ... The tracker had now gone down on his haunches. Moving the short distance to him with my rifle at the ready and as I started to ask what he thought, he took off down the hill as if the devil himself had intervened, narrowly missing taking me with him. The OC was not so lucky, being bowled over by the fleeing man. As I looked up from the empty space where the tracker had been, I noticed a terrorist some ten metres from me.[34]

The importance of tracking in the initial counter-insurgency operations of 1967 and 1968 prompted the Rhodesian Army to develop its own expertise in this field. In 1968 white National Parks wildlife officers, who did obligatory military call ups as conventional infantry or as members of Savory's TCU, refused to continue doing additional tracking for the army without compensation and were awarded danger pay and a cash bonus for successful contact.[35] This influenced the army to begin expanding its pool of competent trackers. According to former SAS commander Brian Robinson, the National Parks and professional hunters employed the only known trackers but 'They believed their task was tracking game and not becoming involved with men who fired back. Soldier trackers were as rare as hen's teeth'.[36] During the late 1960s some Rhodesian army officers and a few non-commissioned officers attended a National Parks tracking course at Wankie National Park. The army's School of Infantry began offering its own tracking courses in the late 1960s and in April 1970 formally established the Rhodesian Army Combat Tracking Wing at Kariba as it was a wilderness area with a good mixture of terrain and was close to insurgent infiltration routes. The first Tracking Wing instructors were some of the best trackers from the SAS. Founding Tracking Wing commander Robinson, a member of the short-lived GATU who had been briefly attached to the British SAS, and Savory developed the concept of a four man tracking team that moved 100 or 200 meters in advance of a larger infantry unit which would be called forward when the enemy was discovered, and tactical drills for a tracker team to use when trying to recover a lost trail. Tracking Wing also adopted TCU practises such as the wearing of shorts and plain soled boots specially made by Bata Shoe Company. Robinson had considerable freedom in writing training manuals as no one else in the army hierarchy was familiar with tracking. Given the use of the Kariba area by the Air Force and BSAP in its experiments with controlling tracker dogs from helicopters, the Tracking Wing tried to use a dog on a long lead to track at night when visual tracking did not work. However, the trials were abandoned when the dogs unexpectedly led their handlers toward wildlife including a pride of lion.[37] Like many Rhodesian officers, Robinson came to think that 'Dogs do in fact have their day but in the main it is too hot, too cold, too dry, too wet or too something. The handler ends up carrying the dog like a cape'.[38] Eventually, Tracking Wing utilized skilled and experienced instructors from other units including National Parks WOs called up for military service.[39] The wing's staff also included some black civilian army employees called magojas (cookers of meat) who, under the supervision of a different candidate every day, would set off from camp in the early morning to leave spoor for the students to follow. The wing's tracking courses were divided into three difficulty levels: basic for beginners, intermediate for students who showed potential and advanced for those who could become really skilled. At times South African Special Forces operators and Portuguese soldiers (see below) took part in the training. Furthermore, the wing offered bush survival courses for other Rhodesian army and air force personnel. The training area around Lake Kariba was richly populated by wildlife including dangerous predators. In April 1972 Lieutenant Al Tourle, the Tracking Wing's commander and a highly respected former sergeant major in the RLI, was attacked by a lion and mortally wounded.[40] Rhodesian infantry officer candidates took a tracking and survival course at Kariba where, as described by RLI soldier Tim Bax who attended in the early 1970s, 'we learned the art of tracking human quarry through the bush. More important, we learned the art of anti-tracking to prevent terrorists from tracking and killing us'.[41] Like other military specialists, soldiers who had completed the advanced tracking course at Kariba received additional pay though the amount was lower for African personnel.[42]

By 1968 the all-white RLI had developed its own tracker combat teams. Some white Rhodesian soldiers had learned tracking while growing up in rural areas such as C. J. Skeepers who joined in the late 1960s and explained that 'I grew up on my dad's farm in Mangula. The bush was my life. When the first European farmers were killed by the gooks in '66, my dad told me the Army would need people like me to help track down the terrorist gangs'.[43] As a fairly new unit, the military culture of the RLI encouraged experimentation by sub-units and the most effective methods were subsequently adopted by all. In the late 1960s RLI officers were sceptical when told by their RAR counter-parts, with their experience in 1950s Malaya, that patrols of platoon or company size were needed to effectively engage insurgents. The RLI considered those tactics too slow and cumbersome. At the end of July 1968, while pursuing a group of 50 ZANU insurgents who had crossed the Zambezi River at Mpata Gorge, Rhodesian Air Force pilot Peter Petter-Bowyer experimented with using a helicopter to leapfrog small teams of RLI trackers toward possible water sources that the insurgents were likely heading for as it was the dry season. While the helicopter-borne trackers made up seven days lost time in a few hours, they ultimately landed too close to the insurgents who heard the sound of the helicopter engines and escaped into neighbouring Mozambique. From this, the RLI developed a standard tracking practice of using helicopters to gain ground on insurgents. One tracker team remained following the spoor while another was flown forward to speculatively search for the same line of tracks though they tried to avoid landing too close to the suspected enemy who might then scatter or set an ambush. Sometimes the forward tracking team sent some trackers to back-track the trail they were following to determine if it indeed had been made by the same or another insurgent group. Another tactic meant to box-in fleeing insurgents was for helicopters to deposit small teams ahead of trackers to ambush or contain fleeing guerrillas. These procedures worked exceptionally well in the semi-open and little inhabited Zambezi Valley during the first phase of the Rhodesian War. There was space for helicopters to land and any unfamiliar tracks were likely those of insurgents.[44]

Fighting insurgents in nearby Angola and Mozambique, Portuguese security forces were particularly interested in their Rhodesian allies' emergent tracking skills. Most Portuguese troops were conscripts from Europe with little knowledge of the African bush or interest in the war. However, there were several exceptions that demonstrated the potential value of tracking to Portuguese commanders. In 1965 Portuguese forces in southeastern Angola began employing local Bushmen organized into a growing formation called 'flechas' (arrows) who conducted their own reconnaissance and harassment patrols along the Zambian border and tracked insurgents for army units. While the Portuguese sought to mobilize the legendary tracking skill of the Bushmen and exploit historic tensions between this marginalized minority and the majority African population, they eventually opened the flechas to former insurgents. Angolan Bushmen joined the Portuguese military because they felt coerced, it gave them access to food and other goods, it gave them relative authority and status, and they and their families were offered protection in the worsening conflict. Some worried that the Portuguese were exaggerating their tracking abilities and that in taking sides in the war they would be seen as traitors by the nationalist movements.[45] In Mozambique, Daniel Roxo, a Portuguese hunter in the territory since the 1950s, formed a private militia from his African trackers and other employees which hunted FRELIMO fighters in Niassa Province and reputedly killed more of them than the regular Portuguese forces.[46] In 1968 and 1969 Rhodesian and Portuguese military, police and intelligence officials conducted secret coordination meetings in Mozambique. It was there that the Rhodesians promised to assist the Portuguese in tracker training. At one of the 1968 meetings the Rhodesians pointed out that Portuguese forces in Mozambique's Tete province were making excessive use of helicopters to transport troops when they should have been focusing more on ground operations such as patrolling, tracking and ambush.[47] Ron Reid Daly, the first Rhodesian officer to advise the Portuguese military in Mozambique in 1967, found them militarily incompetent. They never reconnoitred a target prior to attack, rejected the idea of long duration patrols, surrendered the country to the guerrillas during the night and never conducted follow-up tracking after an insurgent contact.[48] During the late 1960s and early 1970s Portuguese officers visited Rhodesia to observe tracking demonstrations, and Portuguese troops attended Tracking Wing courses. Among the Portuguese students at Kariba was the legendary Daniel Roxo who failed to impress Rhodesian instructors. In 1970 a Rhodesian SAS team went to northern Angola to conduct a tracking course for Portuguese troops who did not seem to take the matter very seriously. Robinson, who led the SAS team, flew in a helicopter above the operational area and observed numerous and well-used possible insurgent trails that no one seemed interested in investigating. Most cooperation between Rhodesian and Portuguese security forces took place in Mozambique as the two territories shared a border. The Portuguese in Mozambique had been late in mobilizing African military manpower and were having trouble containing FRELIMO which had staging areas in southern Tanzania.[49] Beginning in 1967 or 1968, Rhodesian SAS tracking teams assisted Portuguese forces during operations in Mozambique's Tete province where FRELIMOwas opening a new front and which bordered on Zambia and northeast Rhodesia. According to SAS historian Cole, 'Often to the horror of the SAS trackers closing in on fresh spoor for them, the Portuguese would break open beer cans and start drinking ... or bang tins or planks to deliberately frighten the enemy away.'[50] Reid Daly later explained that since the SAS teams were considered elite Special Forces, RLI tracker teams were sent to Mozambique to show the Portuguese what ordinary young soldiers could accomplish with training and leadership. In January 1969 a RLI tracking team, including Pete Clemence who had learned tracking from indigenous people while growing up in Botswana, with Portuguese troops in Tete found the first evidence proving that FRELIMO insurgents had crossed south of the Zambezi River. This represented a major development as up to that point they had been confined to northern Mozambique but now threatened Rhodesia's access to the Mozambican ports. In November 1970 another RLI tracking team with a company of Portuguese paratroopers followed a FRELIMO group for five days until they discovered a base and major arms cache. While the Rhodesians influenced the Portuguese to establish their own Combat Tracking Special Groups in 1970 and Portuguese army trackers began to deploy, it was too little and too late to have any significant impact on counter-insurgency in Mozambique.[51]

Don Price, an RLI officer involved in operations with the Portuguese, wrote 'the single most important thing we learned while in Mozambique was tracking and just how essential the role of trackers was in the success of closing and killing the enemy. It was obvious that it was important to train up our own trackers.'[52] It was in Mozambique that the SAS operators passed along their tracking skills to the RLI soldiers. In 1969 the RLI ran its own two week tracking course in Wankie National Park with Price and Willie de Beer, a retired RLI officer who had entered the National Parks Department, and African game scouts teaching the white troops subjects such as basic tracking, tracking in teams, bush survival, animal habits and bush shooting. These courses continued during the early 1970s and were run in different environments such as the shore of Lake Kariba, the desert conditions of the

Botswana border in Wankie National Park, the Mopani woodland near Victoria Falls and in the operational area of the Zambezi Valley. They eventually served to prepare RLI soldiers for attending Army Tracking Wing courses at Kariba.[53] The success of RLI trackers in Mozambique and the unit's own tracker training led to the formation of a dedicated RLI Tracking Troop in 1971 which was initially under battalion headquarters and dispatched tracking teams to different Commandos (companies). In 1972 Tracking Troop was incorporated into RLI Support Commando and in 1976, with the addition of other conventional warfare duties, it was renamed Reconnaissance Troop.[54] By this time, the development of the RLI into an air mobile reaction unit and the transfer of its pioneer trackers to other units such as the Selous Scouts and SAS meant that 'the RLI tracker teams dissipated'.[55]

Other Rhodesian Army units approached tracking differently. Rhodesia's other regular infantry battalion, the primarily black RAR, did not centralize its trackers and stuck more closely to methods that had been practised in Malaya. With a core of experienced trackers and many soldiers who had grown up herding and hunting in rural areas, the unit conducted its own basic tracking training, sent some soldiers on the Kariba course and spread tracking specialists among rifle companies.[56] In 1968 Sergeant Laurie Ryan, who had excelled in field craft training and eventually took an army tracking course, joined the territorial 4 Rhodesia Regiment (4 RR) based around the eastern town of Umtali on the Mozambique border. He quickly played a central role in that unit forming a specialist tracker team.[57] Despite training many soldiers as trackers, the expansion of the insurgency meant that the Rhodesian Army including the RLI and RAR continued to employ PATU, civilian and National Parks trackers until the end of the war.[58]

Early insurgent responses to Rhodesian security force tracking were mixed. Although nationalist guerrillas appeared to know little about tracking or anti-tracking during the first phase of the war, there were indications that they were learning. In 1968 Rhodesian soldiers pursuing ZAPU infiltrators noticed that they practised effective anti-tracking including walking abreast in a widely spread line which minimized their spoor and crossing a dirt road by laying items of clothing on it as stepping stones to avoid leaving footprints and then removing the cloths. Similar observations were made about ZAPU insurgents in 1970 during operations Birch, Teak and Granite. However, poor anti-tracking by a ZANU group which crossed the Zambezi River in August 1968 led to its quick location and neutralization by an RAR company.[59]

Expansion of the War

From 1972 ZANLA, based in Mozambique, penetrated northeastern and then eastern Rhodesia with small units that focused on politicizing rural communities and tried to avoid Rhodesian forces. Insurgent activity increased during the rainy season as the growing bush provided more cover though moving after rainfall usually left more clearly visible tracks. In this new phase of the war, given previous experience and perhaps taking lessons from FRELIMO, ZANLA began routinely using anti-tracking precautions. It greatly favoured the practice of 'bombshelling' which meant that when a unit was discovered by security forces it would split into small groups or individuals who fled in many different directions to prevent trackers from following them all. The insurgents would then meet at a prearranged location. Other popular anti-tracking methods included walking in streams or on popular footpaths, changing footwear or going barefoot to alter footprints, and having sympathizers brush away their tracks or drive livestock over them.[60] In 1975 Agrippah Mutambara went through ZANLA's basic guerrilla training in Mozambique where he learned how to create the illusion that he had disappeared during an engagement with the enemy by crawling and using cover, and rudimentary anti-tracking such as walking backwards while crossing a dirt road, path or river.[61] A little later, as an instructor, he designed an advanced training program for a special commando unit that learned bush survival, tracking, anti-tracking and animal behavior in a Mozambique game reserve. Mutambara vaguely alludes to the presence of specialist instructors who may well have been Mozambique game-keepers or professional hunters or FRELIMO veterans. Speaking of the students on this advanced course, Mutambara maintains that:

> They acquired a thorough knowledge of tracking skills which, together with the chameleon skills of camouflage and concealment, turned them into an invisible rebel force against the Rhodesian regime. The terrain which they traversed became the main source of their operational intelligence. They were trained to glean intelligence from foot prints and other disturbances on the ground, and to determine how long ago they could have occurred. The foliage too was an important source of intelligence. If there were bent or broken twigs or grass, they needed to decipher whether they were caused by animals or humans, which direction the animal or human was going, whether they were broken in a rushed or casual manner, and the approximate time the damage occurred, as well as the numbers involved. Interpreting accurately the reactions of birds and animals and the different sounds and warnings they make when sensing lurking danger, or just communicating with each other, needed special training and expert observation. Mastering such skills turns birds and animals into useful and dependable allies against an enemy. We thus taught our cadres to be friends of, and friendly with, their environment ... the environment had to be utilized to protect and conceal every movement from the prying eyes of the enemy, to warn of imminent danger, and to secretly lead to an enemy. We trained our cadres to use the environment to their advantage to spring surprise attacks against the enemy. But the environment, can also be like a tame ass. Friend and foe alike can ride it without resistance. Being at one with the environment does not preclude the enemy from being the same. The cadres had to be alert and vigilant at all times in order to outwit the enemy.[62]

Well-known Rhodesian officer Reid Daly explained that in response to the first ZANLA attack on a white farm in northeastern Rhodesia in December 1972, security force units 'deployed immediately with SAS trackers. But apart from one small contact there was nothing: tracks simply dissolved into the tracks of the local inhabitants or, in some cases, were swept away by the locals with branches'.[63] Anthony Trethowan, a veteran of BSAP Ground Coverage, clarifies, 'tracking, ambushing, follow-ups etc. – were simply not working as they had in the earlier days of the bush war in the sparsely populated Zambezi Valley: the guerrillas were now mixing with the povo, the local peasants, in the heavily populated TTLs'.[64] SAS historian Cole believes that Rhodesian trackers who had worked with the Portuguese in Mozambique observed problems that would soon apply to the insurgency in their own country:

> They aimed to show the Portuguese both the tracking concept and the joint operations concept of counter-insurgency warfare ... Yet when it came to tracking, it was the Rhodesians who were to learn the biggest lessons. Up until then, they had viewed tracking as the magic ingredient for success. Their own victories had been in the remote, unpopulated regions of the Zambezi Valley where, if there was no interference from the rain, tracks would remain for anything up to a week. They soon found that tracking was an entirely different proposition in a heavily populated area where it was pouring rain the whole time. It was the first indication to the Rhodesians that should things go wrong and the war be taken to the population, they too would have similar problems. They were not to know it then, but by the time their own war got into full swing, the need for tracking would virtually fall away. There would be so many enemy running around Rhodesia that there would be no need to track them down; and there would not be enough trackers to do the job anyway.[65]

The war was further expanded in the late 1970s when ZIPRA, ZAPU's Soviet supplied armed wing, established staging areas in Botswana to infiltrate arid southwestern Rhodesia though they were more inclined to combat security forces than mobilize the masses. Duncan MacArthur, a Rhodesia Regiment tracker during the late 1970s, described the challenges of tracking in this area:

> Tracking in the Beit Bridge area required a great deal of common sense and experience, because in most cases the ground was far too hard and hostile so you were lucky to get spoor confirmation every five kilometres. The best and easiest way to track in this area was with the sun behind you, providing a good image of the ground that your quarry has just passed through. Most of the time operational tracking is a mind game – you get into the habit of placing yourself in the enemy's shoes and figuring out what you would do if you were being pursued and tracked.[66]

While the Rhodesian security forces responded to the insurgency's expansion by initiating new methods and units, the emphasis on tracking continued as part of the overall focus on killing as many guerrillas as possible. In the early 1970s, Rhodesia began building a 25 meter wide cordon sanitaire along the Mozambique border which was fenced on both sides, cleared of vegetation by chemical defoliants, and seeded with landmines and eventually electronic warning devices. Following the spread of ZANLA infiltration, it was first constructed in the northeast and extended south. A dirt road along the Rhodesian side of the inner fence facilitated tracking. According to former Rhodesian intelligence agent Henrik Ellert, 'The most effective mechanism adopted by the Rhodesians to detect early guerrilla crossing came from the daily spooring patrol which traversed given sections of the border dragging behind their vehicles a wooden framework with rubber tyres. The drag smoothed the surface of the dusty roads and the early morning patrol could easily detect any disturbances'.[67] Although security forces employed the now standard helicopter leapfrogging technique to hasten tracking of insurgents who had crossed the border barrier, there were insufficient resources to patrol its entire length and it was too narrow which meant it 'never proved a serious obstacle to guerrilla infiltrations'.[68] Rural people were herded into protected villages (PVs) to inhibit their support for insurgents and channel guerrillas into uninhabited 'no-go' areas where they could be tracked more easily which was enhanced by bulldozing sections of bush.[69] The squalid PVs also became scenes of tracking and anti-tracking as insurgents, after they had cut a hole in the fence to gain access to food or recruits, used branches to brush away their tracks to avoid showing the daily perimeter patrols the direction they had departed.[70]

In 1974 the Rhodesia Army Tracking Wing and 90 strong TCU were absorbed by the new Selous Scouts which used tracking as a cover for its primary and covert mission of utilizing captured and turned insurgents and black security force personnel to infiltrate and destroy guerrilla groups in 'pseudo-operations'. Lieutenant Colonel Ron Reid Daly, commanding officer of the Selous Scouts, had the Tracking Wing standardize and intensify its programs and placed it in charge of the former TCU reaction teams. White and black regulars and white territorials (there were no black territorials) who passed the Selous Scouts' gruelling two week selection course continued with a two week tracking course but then the white territorials were sent home and the regulars embarked on covert training in how to impersonate guerrillas. Although the white territorials of the old TCU were unsuitable for pseudo-operations against black insurgents, their continued work as army trackers helped maintain the unit's cover and in the late 1970s they formed conventional motorized columns for cross-border raids.[71] According to Major General Archer Bruce Campling, a veteran of Malaya and a Rhodesian Army brigade commander in the late 1970s, 'the Selous Scouts continued to train and deploy trackers until the end of the war but this became very much of a secondary role and it was neglected to the detriment of the rest of the Army effort'.[72] Indeed, recruitment for the Selous Scouts drained the SAS, RLI and RAR, the core units of the regular army, of many of its best personnel including the few really skilled trackers. It also undermined the Kariba Tracking Wing which became the Scouts' Training Troop and focused on pseudo-operations. Other units hesitated to send their personnel there for tracker training as they might be enticed into joining the increasingly celebrated Scouts. It seems that the staff of Training Troop may have put their tracking and hunting skills to other uses as some were accused of poaching in the Kariba wilderness area and illegally selling products from endangered species such as ivory and rhino horn.[73]

Many incidents that occurred during the second half of the 1970s illustrate the increasing frustrations of Rhodesian security forces trying to use tracking to locate and engage ever more active and numerous insurgents who employed effective anti-tracking measures and made for crowded TTLs when being followed. After a dozen ZANLA fighters erected a road-bloc on the Beit Bridge to Fort Victoria road and killed several white motorists in April 1976, two Selous Scouts tracking teams under Lieutenant Ant White were deployed to the scene where they followed the insurgents for over 100 kilometres over four days sometimes using helicopters to leapfrog. However, the trackers lost the insurgents' spoor when they entered the populated Sengwe TTL where their footprints merged with those of numerous civilians.[74] The war diary of RAR company commander Major Andre Dennison, a British veteran of several counter-insurgency campaigns, reveals similar frustrations. In February 1977 his company deployed by foot to check reports of insurgents in a hilly area bordering a TTL and followed the fresh tracks of eight suspected guerrillas who split up with some entering 'Angus Ranch area and the spoor was lost. The rest cut back into the TTL and spoor was finally lost in the area of Chiremwaremwa School and Matsai Business Centre. A very footsore call-sign returned to Mashoko'. At the end of December of the same year Dennison, describing an air mobile reaction to an insurgent ambush of military vehicles, wrote 'Trackers were flown in ... Fire Force arrived on the scene, but were totally unhelpful, and the trackers performed their usual trick of losing spoor'.[75] Rhodesian Special Branch officer Ed Bird recalled an insurgent night raid on a white-owned farm in the southern part of the country in January 1977 which was driven off by a police patrol. The next morning an army tracking team arrived at the scene and began to pursue the fleeing insurgents but 'The tracks headed in the direction of the Diti Tribal Trust Land where they were soon lost'.[76] Based on the BSAP Incident Log for the Beitbridge area in the late 1970s, Bird's account of security force operations in the south lists numerous instances of the success of insurgent anti-tracking. After an insurgent attack on a protected village in late April 1978, 'C Company 1RAR conducted a first-light follow-up but, due to anti-tracking methods employed by the terrorists, tracks were lost after about five kilometres'.[77] In October 1979 'Elements of 1 Indep Company were deployed and located spoor of fifteen terrorists which they followed for three kilometres before losing it to the enemy's anti-tracking skills'.[78] Lieutenant Noel Smith, in January 1977, was leading a 1 Rhodesia Regiment (1 RR) ground patrol in the Masoso Tribal Trust Land in north-eastern Rhodesia:

> I came across the tracks of approximately 15 terrs heading in a south-westerly direction. I radioed base but was informed that Fireforce was unavailable and was ordered to continue on the tracks with my men. We began an exhausting six-hour follow-up, expecting an ambush at any moment, before losing the tracks when they intermin-

gled with cattle spoor. This appeared to have been a deliberate anti-tracking tactic and it was obvious the locals had alerted them to our presence. We attempted to relocate the tracks but finally gave up when a very heavy rain shower fell in the late afternoon. It had been a frustrating 11 hours.[79]

With tracker training and reaction teams taken over by the Selous Scouts who were busy with pseudo-operations, the Rhodesian Army formed a number of other specialized tracking units and sub-units in an effort to cope with the expanding insurgency of the 1970s. In 1975 the Rhodesian Army created a horse mounted infantry unit called the Grey's Scouts that patrolled the Zambian and Mozambican borders for signs of insurgent infiltration. Attempts to monitor the frontier with electronic devises had failed for technical reasons and motor vehicles proved unsuitable for the rough terrain. By 1978 the unit consisted of three operational squadrons of regulars and territorials. Most Grey's Scouts were white as horses and equestrian sports were the preserve of white society in Rhodesia. However, the unit contained a few African regulars who had experience with horses as grooms or ranch workers in previous civilian life and efforts were made to recruit some Shangaan, an ethnic group from the remote southeast, who were 'expert and experienced trackers'. According to unit veteran Michael Watson, their main assignment was:

> Observation and location of terrorist tracks, immediate reporting and reaction to sighting of tracks, rapid deployment onto tracks when located, requesting assistance if necessary... using the speed and height of our horses to rapidly follow enemy tracks. Our aim was only to get to engage with the enemy in the quickest possible time, invariably initiated when ambushed.[80]

In 1975 the territorial 4RR, in the context of its border area with Mozambique becoming the scene of increasing insurgent infiltration, was assisted by the SAS in expanding and improving its existing tracker unit which was composed of farmers, game rangers and hunters. Typically, it dispatched a three or four man tracker team to the scene of an insurgent encounter and began tracking until making contact at which time an air mobile 'Fire Force' unit was called to surround and eliminate them. In mid-November 1976 a 4RR tracker team led by Sergeant Laurie Ryan accompanying an RLI Fire Force was responsible for the death of 31 insurgents in the Honde Valley on the Mozambique border which represented the largest number of fatalities inflicted by the security forces in a single engagement up to that time.[81] According to RLI veteran Chris Cocks who participated in the battle, 'We regarded the 4th Batt trackers as some of the best in the Army'.[82] Sergeant Ryan eventually received the Military Forces Commendation (Operational) as 'Through his enthusiasm he has built up the tracker team to be a tremendous asset to the Unit and the Brigade, and has led tracker teams in many contacts which have resulted in success for the security forces'.[83] In

1977 1 RR formed a motorcycle troop, which operated in self-sufficient groups of seven to eight riders, to patrol border and security fences, and react to reports of insurgent activity. After a few months the border patrol duties were reduced as it was found that the riders were vulnerable to ambush and they focused on reaction duties with including tracking with men and dogs. On one occasion a small Selous Scouts tracking team was assisted by these motorcycles in try-ing to catch up to a ZANLA unit weighed down by carrying a large recoilless gun but the spoor was lost when the insurgents drove cattle over their trail and the motorcycles had to withdraw at dusk.[84] In 1978 Major Don Price, a former commander of the Kariba Tracking Wing, was in command of 1 Independent Company based in the Victoria Falls area where information on insurgent activ-ity was difficult to obtain. Without seeking approval, Price formed and trained his own small pseudo team to collect intelligence across the border in Botswana and a Tracking Troop that focused on 'regular patrolling of the gorge area of Victoria Falls to cross-grain for spoor of infiltrating insurgents'.[85] The Rhodesian Air Force also developed tracking skills and by the mid-1970s 'Aerial tracking is now of such a high standard that a skilled pilot can determine whether the blurs in the grass below denote merely a game trail or have been hacked by terrorists to make a path'.[86] The intrepid Petter-Bowyer, during the early 1970s, developed a method for identifying concealed insurgent camps based on aerial observation of trails made by people going to relieve themselves at night which were called 'crap patterns'. Following detection by fixed wing aircraft, helicopters trans-ported trackers to the location of the spoor and were used for leapfrogging.[87]

By the late 1970s the expansion of the Rhodesian security forces, including many white and black urban young men who lacked bush skills, and the escala-tion of the war led to a proportional shortage of trackers. A young white farmer lamented his increasing military call-ups stating 'It's in and out of the army all the time. It makes life extremely difficult for every farmer. Maybe impossible. I mean, how can you track terrs all day and then return and try to farm. You can't ... Its almost impossible'.[88] Although part of covert South African assis-tance to Rhodesia in the late 1970s included the deployment of South African Police combat tracker teams, this was not enough to meet the rapidly expanding requirement for security force trackers.[89]

The greater need for trackers corresponded with the expansion of the insurgency into the southeastern region which was home to the marginalized Shangaan minority already widely known for exceptional tracking abilities. This led to several efforts to mobilize Shangaan trackers in support of Rho-desian security forces. Their employment in the new Grey's Scouts mounted unit has already been mentioned. In early 1977 private game rancher and ter-ritorial Selous Scout Mark Sparrow formed the Civilian African Tracking Unit (CATU) that employed skilled African civilian trackers who, given their age or lack of formal education, were unable to enlist in the security forces. Lieutenant Colonel Henry Dunn, commanding officer of the territorial 8 RR which initially employed Sparrow's trackers, saw the potential of the project and recommended it for formal approval to Joint Operations Commander Lieutenant General Peter Walls. CATU recruited mostly from the marginalized Shangaan of the southeast where Sparrow owned a game ranch. Based at a protected camp, each CATU team consisted of six black trackers under a white 'controller' and were ready to react to a report of insurgent activity. In addition to wages and food, CATU members received cash incentives for each insurgent killed as a result of their tracking. While they were uniformed and armed, CATU trackers were given minimal military training and relied on other units to engage insurgents they discovered. To avoid guerrilla retaliation, their identities were kept secret and they worked outside their home areas. CATU relied on endurance tracking, covering 50 kilometers over two days by foot, to catch their targets or locate insurgent camps, and at times the trackers posed as civilians to collect intelligence.[90] During the late 1970s Bruce Cook of Mazunga Ranch often took his Shangaan trackers, civilian employees at the farm who were also uniformed and armed members of the African Police Reserve, to assist security force units in locating insurgents. These trackers appear to have operated locally and if spoor headed out of their area they passed it over to the regular force trackers such as those from the Selous Scouts. While Cook was an 'outstanding tracker', the Shangaan were much admired by their white superiors and seen as 'blessed with phenomenal bushcraft and tracking skills'.[91] A 1979 effort to use Shangaan civilians to teach Rhodesian soldiers to track was aborted when several instructors were killed in an insurgent attack on the training camp.[92]

An obsession with increasing the insurgent body count prompted renewed experiments, in the late 1970s, with following tracker dogs from helicopters. However, the Air Force found that the nervousness of dog handlers in combat caused their dogs to ignore the scent.[93] Similarly, the Selous Scouts imported a pack of six famously fast Irish Foxhounds, wearing fluorescent jackets, which were set on an insurgent track by an operator who then followed them in a helicopter and would call in an air mobile Fire Force when the enemy was located. Military historian Wood explains that 'the hounds, because of their speed, had a far greater chance of catching up with the terrorists than trackers on foot'.[94] Despite performing incredibly well in trials at Inkomo Barracks, the dogs proved unworkable without direct human control as when employed in the field around Mtoko they ran in the wrong direction when confronted with the meeting point of two insurgent groups.[95] In September 1977 a frustrated Major Dennison wrote in his diary that 'the Fire Force was called away and spent the next few hours following some of Chris Hallamore's fox-hounds in the aftermath of a Scouts' contact. The dogs did not perform as well as had been hoped and at the end of the day two were still adrift'.[96]

The Rhodesian Army and BSAP continued to employ civilian African trackers throughout the 1970s including indigenous minorities and foreigners. During the late 1970s 'a wizened old Bushman' named Maplanka worked as a civilian tracker for the BSAP in Matabeleland. While Maplanka refused to carry a firearm and avoided combat, he was eventually shot to death during a patrol in dense bush when a nervous white policeman mistook him for a guerrilla. According to BSAP veteran Anthony Trethowan:

> Maplanka started off slowly then as the spoor became clearer and a picture formed in his mind he would pick up the pace. Typically Maplanka would be hot on the spoor, going at a steady pace and then stop suddenly. All heads would turn towards him. Without saying a word he would point towards the bush in front of him with his stick and then lie down on his stomach with his face buried in the sand. Literally seconds later a fire fight would ensure. So when Maplanka stopped and pointed it meant we were right on top of the terrorists. He had done his job – now it was our turn.[97]

In 1978 Rhodesian Regiment white soldiers patrolling the Botswana border worked with Julius who 'wasn't an African soldier but an excellent civilian tracker from the Victoria Falls/Botswana area'.[98] Around the same time Duncan MacArthur, commander of 1 Independent Company's Tracker Troop, employed a Namibian tracker named Nelson Tshuma who was involved in discovering a very large insurgent entrenched position.[99]

As the war escalated, the roles of tracker and prey reversed to some extent. Anti-tracking became a priority for the Selous Scouts. As they often located guerrillas by using hilltop observation posts occupied surreptitiously at night, Scouts had to carefully avoid leaving spoor in places that young boys (or mujibas) working for the insurgents checked every morning.[100] At times white Selous Scouts would leave their hill top observation posts at night to secretly confer with black colleagues posing as guerrillas and collecting intelligence among the local population. Former Selous Scout officer Tim Bax describes a typical incident where 'After spending time gleaning whatever intelligence we could, the group (of pseudo-terrorists) would slink back into the night and Bruce and I would return silently to our hill, making sure to anti-track the whole way'.[101] Similarly, small teams of Selous Scouts and SAS operating covertly in neighbouring Zambia, Botswana and Mozambique to gather intelligence on insurgent bases or sabotage infrastructure employed anti-tracking to remain undetected. While planting mines in Zambia, SAS operators wore 'carpet over-boots to avoid leaving the spoor of the soles of our combat boots'.[102] Prior to a raid on a ZANLA camp at Mapai some 120 kilometres inside Mozambique, four Selous Scouts operators reconnoitered the target by crossing the border on a custom built quad cycle that was silently pedaled along a rail line and 'won't leave any tracks to compromise your presence'.[103] In 1979 a Rhodesian Special Branch team assassinated the black manager of a white owned commercial farm in the southern region who was very strongly suspected of aiding guerrillas. The team concealed its involvement by entering and exiting the farm from across the nearby South African border, employing anti-tracking methods while in Rhodesian territory and shooting the man with a captured Soviet-made weapon.[104] Since they usually had considerable experience tracking insurgents at home, Rhodesian operators on external missions were keenly aware that it was almost impossible to anti-track in sandy or wet soil and that walking over hard ground was their best option. The Selous Scouts' practice of using just one or two men on external operations was meant to minimize the change of discovery, including by reducing the number of tracks. Within Mozambique, FRELIMO troops adopted a method of tracking infiltrators to a patch of forest, circling it to look for signs of exit and then shooting into the trees to drive out anyone hiding inside. The Rhodesians were not the only ones to mobilize the tracking skills of marginalized groups as during the 1970s ZAPU employed civilian Bushmen trackers along the Botswana-Rhodesia border and FRELIMO soldiers in Mozambique's Gaza province were often Shangaan trackers.[105]

Independent Zimbabwe and Anti-Poaching

With the negotiated end of the Rhodesian war and the independence of Zimbabwe in 1980, former state and insurgent forces were integrated into a new security force structure. This meant that immediately after independence the Zimbabwe National Army (ZNA) and Zimbabwe Republic Police (ZRP), to some extent, inherited and put to use the tracking expertise of the old Rhodesian forces. In this new dispensation, members of the disbanded Selous Scouts formed the core of a new Zimbabwe Parachute Regiment, the old RLI became the Zimbabwe Commando Regiment and the Rhodesian SAS turned into the Zimbabwean SAS. During the early 1980s some veterans of ZIPRA, frustrated with marginalization and abuse within the new military, rebelled against the increasingly authoritarian ZANU-PF government and took to the bush of rural Matabeleland. Some became involved in a South African supported group called Super-ZAPU that was meant to destabilize the newly independent majority-ruled country and others tried to recreate the old ZAPU network but suffered from lack of leadership and external sponsors. While these 'dissidents', as they were called, easily avoided the ZNA's North Korean trained 5 Brigade which displayed very poor bush craft and focused on terrorizing civilians, they greatly feared the tracking and combat abilities of former Rhodesian units such as the Zimbabwe paratroopers and commandos, and para-military Police Support Unit. Tracking remains an element of ZNA Special Forces training up to today.[106]

While post-colonial Zimbabwe's approach to wildlife conservation differed from Kenya, the recent history of conflict influenced the rise of militarized anti-poaching operations particularly in terms of tracking. In 1980s Zimbabwe elephants were culled and the harvested ivory sold to generate revenue for National Parks, and there were serious attempts to involve indigenous communities in managing the natural environment as a source of renewable natural resources.[107] However, immediately after Zimbabwe's independence there was a dramatic increase in the killing of black rhino – the horn of which is used in making dagger handles in Yemen and aphrodisiacs in Asia – in Zimbabwe's Zambezi Valley by well-armed and organized poachers based across the border in impoverished Zambia and linked to international smuggling networks. Consequently, in 1984 Zimbabwe's National Parks Department launched Operation Stronghold which represented the mounting of small military style anti-poaching patrols which shot suspected poachers on sight. According to Glen Tatham, the chief warden in charge of Operation Stronghold who was called 'General Paton' by his subordinates:

> A prerequisite to the commencement of the anti-poaching operation is to structure strategy, tactics and other efforts along military lines. Combating poachers armed with automatic rifles is akin to an anti-guerrilla type warfare. In the first instance the area has to be secured by intensifying ground coverage and establishing strategic forward basecamps.[108]

A sport hunting journalist wrote that 'Black and white scouts, whose fighting experience in the bush goes back to opposite sides of the Rhodesian war for independence, are combining radio equipment, a helicopter, training and motivation with a clear cut objective. Patrols can be on the trail of an armed gang in hours instead of days – the consequences are often fatal'.[109] Operation Stronghold depended on personnel with military experience as it was directed by white senior rangers who had fought in the bush war of the 1970s, the veteran black rangers had tracked for the Rhodesian forces and some former insurgents were specially recruited probably for their anti-tracking expertise. However, junior black personnel lacked sufficient military training and were reluctant to engage poachers. In 1989, in response to the Zimbabwe police charging several rangers with the murder of suspected poachers, the government passed a law providing immunity from prosecution to members of anti-poaching patrols. Some 170 alleged poachers were killed between 1984 and 1993. The patrols were augmented by Zimbabwe police and the operation was funded by the United States Agency for International Development (USAID) and international conservationist organizations. In 1987 Zimbabwe's National Parks Department established nine elite anti-poaching units; one for each of Zimbabwe's provinces plus one to assist provincial units during emergencies. While Operation Stronghold was expanded to other parts of the country, black rhinos were relocated to intensive protection zones and eventually many were dehorned to remove the reason for hunting them. In 1993 the state launched Operation Save Our Heritage which mobilized military and special police units in support of National Parks anti-poaching activities. Gonarezhou National Park became a 'frozen area' in which rangers and soldiers, reminiscent of the previous Selous Scouts scandal, were eventually accused of complicity in poaching.[110] Given these failures, during the 1990s National Parks began moving rhinos to privately owned land converted from cattle ranches to nature conservancies that would derive revenue from tourists and hunting of other species. Kenneth Manyangadze, who as a national parks game scout during the late 1970s had tracked for the Rhodesian forces and took a leading role in Operation Stronghold, resigned from National Parks in 1988 because of low pay and declining morale. In 1992 he became chief scout at the Save Valley Conservancy which is dedicated to preserving rhino.[111] During the 2000s the collapse of the Zimbabwe economy, related to the occupation of white owned commercial farms and Western sanctions, led to a dramatic rise in poaching which the National Parks Department struggled to counter given lack of resources. Around 2010 Pete Clemence, known as one of the best trackers in the old Rhodesian Army who was rumored to have resigned from the Selous Scouts over poaching within the unit and went on to become Zimbabwe's Conservationist of the Year during the mid-1990s, and his son Bryce Clemence formed Aggressive Tracking Specialists (ATS) which began training personnel in each of the five Intensive Protection Zones where rhinos were fitted with radio tracking devices and constantly monitored by game rangers. ATS training was funded by Save Foundation of Australia which is dedicated to rhino preservation. The training program offered basic, intermediate and advanced courses focusing on tracking in different environments and seasons, and associated subjects such as navigation, bush craft, shooting, coordination with aircraft, close combat techniques, camouflage and concealment, observation and operational planning. The four man tracking procedure taught by ATS was the same as that developed at the Rhodesian Army Tracking Wing in the early 1970s. The 55 rangers who graduated from the ATS advanced course formed a 'new elite unit of AGGRESSIVE COMBAT TRACKERS within Zimbabwe's Parks and Wildlife Management Authority'.[112] Tracking in anti-poaching operations is similar to counter-insurgency tracking though some slightly different approaches have developed in Zimbabwe. Since poachers frequently practice effective anti-tracking techniques such as tying goat skins or foam rubber to the soles of their feet, anti-poaching patrols that observe signs of a poacher presence often try to locate the more easily tracked rhino that is believed to be in danger and then set an ambush for the poachers as they stalk their prey.[113]

Conclusion

Given their operational environment which was generally unfavourable to vehicles, Rhodesian trackers attempted to catch up to fleeing insurgents by either moving fast on foot or leapfrogging in helicopters. Rhodesian counter-insurgency tracking was at its most effective in the late 1960s when the war was fought in the favourable environmental conditions of the remote and sparsely populated Zambezi Valley and the novice insurgents knew little about tracking or anti-tracking. When the war shifted to the rougher terrain of the more populated east in the 1970s, and insurgents began practising effective anti-tracking techniques, the recruiting and training of Rhodesian security force trackers was expanded and new methods such as pseudo-operations and air mobile Fire Forces were devised to locate and kill insurgents. The Rhodesians had difficulty incorporating tracking into their security force structure. The only dedicated tracking unit was the TCU which consisted of a small number of white part-time soldiers, the elite all-white SAS served as a pool of tracking teams but this distracted it from covert operations and the much romanticized Selous Scouts were really a pseudo-terrorist unit that used tracking as a cover and stole the best trackers from other units. With the exception of the scattered employment of a few trackers from small and marginalized groups like the Bushmen and Shangaan, Rhodesian counter-insurgency planners did not systematically mobilize the tracking skills of indigenous communities and most Rhodesian trackers during the war were from the white minority. Indeed, this has engendered a certain pride in white veterans of the Rhodesian forces as Alexandre Binda states that 'On the whole the weight of day-to-day operational tracking fell on acculturated and trained white personnel'.[114] This reflected the long standing white Rhodesian anxiety over arming African troops that delayed expansion of the RAR until the late 1970s. It may have also been linked to the confident frontiersman identity of white Rhodesians who imagined themselves as masters of the African wilderness like their nineteenth century heroes Selous, Burnham and Baden-Powell.

5 SOUTH WEST AFRICA (NAMIBIA), 1966–90

German colonization in South West Africa began in 1884 with Berlin's sudden entry into the European race for African territory. Unlike its other colonies in tropical West and East Africa, the Germans viewed the arid grasslands between the Kalahari and Namib deserts as a potential place of European settlement along the lines of the neighbouring Boer (Afrikaner) republics and British colonies that would eventually make up South Africa. During the 1890s the Germans consolidated their control through a divide-and-rule strategy whereby they made an alliance with the Herero and crushed Nama resistance. However, in 1904 the Herero rebelled against German land alienation and oppressive debt collection. After defeating the Herero at a major battle on the Waterberg Plateau, German military commander Lothar von Trotha ordered the extermination of these people who were driven into the Kalahari Desert and kept away from waterholes. The Nama joined the rebellion and were subjected to the same treatment. In what became the first genocide of the twentieth century, Herero and Nama were eventually herded into concentration camps where, until 1907, they were used as slave labour and forced to undergo deadly medical experiments. The subsequent shortage of African labour for settler ranches and regional mines impelled the Germans to extend their control over the Ovambo and other peoples in the north close to the border of Portuguese Angola. Like all the German colonies in Africa, South West Africa was invaded by the Allied powers during the First World War. For imperial Britain, the immediate objective of the campaign was to deny the use of South West Africa's south Atlantic ports to German naval raiders. For the government of South Africa, a new self-governing British dominion that carried out the invasion of its German neighbour, the campaign held out the promise of territorial expansion. Although delayed by a Boer rebellion at home, the South African invasion force used armoured cars and aircraft to quickly outmanoeuvre the Germans who surrendered in 1915. In 1917 a South African expedition solidified control of the north by defeating the forces of Mandume Ya Ndemufayo, ruler of the Kwanyama Ovambo. After the First World War Germany was stripped of its colonies and South West Africa became a South African administered mandate of the League of Nations. While mandates were theoretically meant to be prepared for self-government in the very distant future, South Africa ruled South West Africa as a new province and imported poor Afrikaner settlers. In 1946 South Africa refused to transfer its mandate to the new United Nations (UN) which wanted to impose more rigorous international supervision of the administration of South West Africa. South Africa did not annex the territory but its white minority enjoyed some representation in the South African parliament. Following the 1948 election of the Nationalist Party in South Africa, apartheid policies were extended to South West Africa and in 1959 eleven black protestors were shot dead by police during the forced removal of their community from the territorial capital of Windhoek. African nationalist movements were formed such as the South West African National Union (SWANU) in 1959 and the South West African People Organization (SWAPO) in 1960 which sought UN trusteeship and eventual independence. Harassed and imprisoned by South African security forces, nationalists began to flee the country to seek assistance from sympathetic nationalist leaders in other parts of Africa that were becoming independent. International revulsion over apartheid and the rapid European decolonization of Africa impelled the UN, in 1966, to revoke South Africa's mandate over South West Africa.[1]

In 1962 exiled SWAPO activists, under founding leader Sam Nujoma, formed an armed wing called the People's Liberation Army of Namibia (PLAN) to fight South African occupation. During these early years of exile SWAPO was supported by newly independent Tanganyika (soon to be Tanzania) and Zambia. SWANU eventually lost international support and withered as its hesitance to engage in armed struggle meant that the OAU and UN recognized SWAPO as the legitimate representative of the Namibian people. Furthermore, the Sino-Soviet split played a role in the demise of SWANU which received fleeting support from communist China while SWAPO accepted more robust sponsorship from the Soviet Union which also supported the prominent liberation movements ZAPU and the South African ANC.

The geography of South West Africa presented SWAPO/PLAN with some problems in infiltrating the territory and launching a guerrilla campaign. Establishing staging areas in neighbouring countries was difficult. In the north, Angola represented enemy territory as it was ruled by colonial Portugal which had allied with apartheid South Africa. PLAN cooperated with Angolan insurgents from the Movement for the Popular Liberation of Angola (MPLA) and the Union for the Total Independence of Angola (UNITA) who were fighting the Portuguese but it was unable to establish permanent bases in southern Angola. The western frontier South West Africa was even worse. The vast and inhospitable Kalahari Desert, into which the Herero rebels had been driven in 1904, dominated the border with Bechuanaland which was a British colony until it achieved independence in 1966 as Botswana where the poor and conservative black government could not afford to alienate white ruled South Africa and Rhodesia upon which it was economically dependent. The border with Zambia, SWAPO's only friend in the area, was very narrow and located at the end of the long and thin Caprivi Strip which was geographically remote from the main body of the territory. Even when insurgents made it into South West Africa, they still had to contend with a number of limitations. Generally, guerrilla fighters rely on natural cover such as forest to conceal their presence from a powerful state force and a local population from whom to derive support and recruits. Both elements were often missing in large parts of South West Africa. While most of South West Africa's population lived in the north along the Angolan border, the important economic and state infrastructure that represented potential targets for hit-and-run attacks lay to the south. The heartland of SWAPO political support was Ovamboland which was a flat area in the centre of the northern region containing over half the entire country's population. However, most of this support was located in western Ovamboland which was a semi-desert characterized by scattered, low vegetation and termite mounds that did not constitute much cover for insurgents. The people of eastern Ovamboland, which included the Kwanyama, were generally less sympathetic to the movement though they lived in a more thickly forested area well suited for guerrilla operations. Other parts of the far north were also problematic as the Northern Namib Desert in the extreme west near the Atlantic coast consisted of inhospitable sand dunes, Kaokoland in the west was mountainous, exposed and thinly populated, and eastern Kavango was mostly underpopulated, sandy and open. As will be discussed below, once the Portuguese had left Africa, South Africa's alliance with UNITA which was based in south-east Angola prevented SWAPO from moving across that border into adjacent Kavango during the late 1970s and into the 1980s. Forming a buffer between the densely populated African communities of Ovamboland in the north and the white ranching area of central South West Africa, Etosha National Park represented a vast uninhabited wilderness at the middle of which was the completely parched, flat and treeless Etosha Pan. If insurgents, by some miracle, arrived in the white farming area they would fail to find much thick bush in which to hide or African communities from whom to gain assistance.

In 1965 small groups of PLAN insurgents began to infiltrate South West Africa from western Zambia though many were intercepted by Portuguese security forces while trying to cross southeastern Angola. Some made it through and established a camp, which included underground shelters and eventually housed 50 men, at Ongulumbashe in northwest Ovamboland which was meant to serve as a base for the spread of a guerrilla campaign throughout the northern rural areas. However, South African security forces discovered the camp and in late August 1966 it was destroyed by a detachment of South African police and paratroopers transported in helicopters. A limited insurgency commenced. In September 1966

PLAN attacked the border town of Oshikango and in March 1967 it ambushed a police patrol in Western Caprivi but in both cases most of the insurgents were killed or captured. PLAN commander-in-chief Tobias Hainyeko, in May 1967, was killed in a skirmish with police in Eastern Caprivi. In October 1968 two large PLAN groups moved across the border but by year end 178 had been killed or captured. Given these losses, during the late 1960s and early 1970s PLAN focused primarily on infiltrating small groups into the Caprivi Strip, targeted local headmen and chiefs who worked with the South African administration, and planted landmines on dirt roads. With labour unrest spreading across the country in 1972, PLAN accelerated its activities over the next two years which led to the deployment of more South African military resources in northern South West Africa.[2]

During the late 1970s and 1980s the insurgency in South West Africa was intimately linked to escalating civil war and South African conventional military intervention in neighbouring Angola. The 1974 military coup in Portugal resulted in the independence of Angola the following year where fighting broke out between MPLA around the capital of Luanda, the National Front for the Liberation of Angola (FNLA) mostly in the north near the Zaire (now Democratic Republic of Congo or DRC) border and UNITA based in the south. It appeared MPLA would seize the Angolan state given its control of the capital, and Cuban and Soviet military backing. This concerned South Africa as Angola would then likely offer support to SWAPO/PLAN and the United States, given the Cold War context, worried about the Soviets gaining access to Angolan oil. South African military intervention began in August 1975 when South African Defence Force (SADF) elements in South West Africa crossed the border to secure an important hydro-electric plant. Around the same time SADF personnel began supplying and training FNLA and UNITA. With Cuban and Soviet supplied armoured vehicles, MPLA conducted a successful offensive and seized much of the country from its local rivals. At the start of October 1975 the South African military, in Operation Savannah, invaded Angola from the south in support of UNITA and quickly advanced north to threaten Luanda. Further north, in late October and early November, FNLA supported by Zairian troops, American sponsored mercenaries, and South African heavy artillery was also trying to take the capital. A Cuban expeditionary force arrived in November and December, crushed the northern FNLA offensive and pushed UNITA and the South Africans back to the south. Embarrassment over public revelations of cooperation with apartheid South Africa led to the withdrawal of American support for FNLA and UNITA, and in March 1976 South African forces withdrew into South West Africa. As the internationally recognized government of Angola, MPLA allowed SWAPO/PLAN to establish training and operational bases in its southern region which quickly increased insurgent activity across the border. In 1977 the South African government reported that during the previous two years

231 insurgents, 33 security force personnel and 53 civilians had been killed in northern South West Africa, and that 300 PLAN insurgents were operating in the territory at any time with 2000 more in Angola and another 1400 in Zambia. The war intensified as engagements, mostly small skirmishes, between PLAN and South African security forces averaged about 100 a month. During the late 1970s South Africa increased its military presence in northern South West Africa including the construction of numerous command and logistical bases. In 1977 and 1978 there were around 6000 to 7000 South African military personnel in the region which was divided into two Military Areas. With its headquarters at the town of Rundu in the east, 1 Military Area (1MA) consisted of Kavango and Caprivi. Ovamboland and Kaokoland fell under the authority of 2 Military Area (2MA) which was commanded from the growing base at Oshakati. 2MA consisted of four infantry battalions along with engineer, armoured car and anti-aircraft elements made up of white national servicemen or reserve Citizen Force/ Commando members on three month rotation in the operational area. It also had a paratrooper company serving as a quick reaction force deployed by helicopter and transport plane. The newly formed 32 Battalion, a large special unit of black Portuguese-speaking Angolans many of whom had been members of the now defunct FNLA, and South African Special Forces (commonly called 'recces') dispatched small patrols across the border to harass SWAPO/PLAN.

By 1978 SWAPO/PLAN and the South African state were competing to regain the initiative in South West Africa. Seeking to undermine a western sponsored scheme for independence, SWAPO began assassinating black political leaders working within the existing system and attempted to create liberated zones in the north. The South Africans realized that they had to strive to do more than respond to the latest insurgent incident and take decisive steps to destroy SWAPO logistics and command. In early May 1978 the South Africans launched Operation Reindeer which represented a three part offensive against SWAPO/PLAN bases in Angola that would open with a large airborne raid on Cassinga some 250 kilometres inside that country and continue with an overland and helicopter attack on PLAN forward operating bases within about 25 kilometres of the border. In the unfolding propaganda war, the South Africans claimed that the 600 people killed at Cassinga were mostly insurgents while SWAPO maintained that they were helpless civilian refugees. In 1977 South Africa had resumed support for UNITA which quickly dominated the south-eastern corner of Angola and by 1980 had ventured into the central region to capture the town of Mavinga which became its main forward base. This distracted the Angolan state and Cuban forces from the border with South West Africa and a desperate MPLA demanded that SWAPO, as a guest in its territory, contribute to the fight against UNITA which was further bolstered by renewed American support in the 1980s. South African raids on SWAPO bases in western Zambia in March 1979 resulted in the group's expulsion from that country. In 1980 the South West African Territorial Force (SWATF) was formed to give the impression that control of the counter-insurgency campaign was being passed to local authorities. However, the SADF retained overall command and seconded personnel to the SWATF which consisted of seven Permanent Force infantry battalions, most oriented around a specific African ethnic group and 27 part-time, multi-racial battalions. The SWATF eventually numbered around 22 000 to 25 000 personnel. Around the same time the military organization of South West Africa was restructured with Sector 10 including Kaokoland and Ovamboland, Sector 20 covering Kavango and western Caprivi, and Sector 70 representing eastern Caprivi. Unlike in Kenya in the 1950s and Rhodesia in the 1970s, South African security forces in South West Africa did not resettle local people into 'protected villages' to try to prevent them from supporting the guerrillas. This was because of the nature of agriculture in northern South West Africa, the Afrikaner historical memory of suffering in British concentration camps during the Second Anglo-Boer War and the potential negative international reaction. Launched in June 1980, South Africa's Operation Sceptic was a sudden mechanized assault by three battle groups on a SWAPO command and logistics complex in southern Angola and represented the first instance of direct combat between the SADF and Angolan army, and the first time SWAPO mechanized units were employed. As a result, SWAPO withdrew its forward operating bases from the border and integrated its logistical system with the Angolan military. Throughout the 1980s the SADF continued to mount ever more ambitious and usually successful conventional military incursions into southern Angola to destroy SWAPO staging areas. These included Operation Protea in late August 1981 which aimed at gaining control of south-central Angola and eliminating the Angolan army's logistical support of SWAPO, and represented the largest South African mechanized operation since the Second World War and its largest deployment of airpower up to that time in the border conflict. In early 1983 PLAN stepped up infiltration efforts by sending 1700 men formed into many small units across the border. In the subsequent fighting, 309 insurgents and 27 security force members were killed.

The success of Operation Askari, a December 1983 offensive by four South African battle groups against SWAPO positions deep in southern Angola, led to the Lusaka Agreement of February 1984 which stipulated that South African forces withdraw from southern Angola and the Angolan government ensure that SWAPO and Cuban units did not occupy that area. By April 1985 the South Africans had moved south of the border but SWAPO, not party to the agreement, quickly returned to the frontier. In late June 1985 South African units again crossed into southern Angola following SWAPO insurgents to their camps which were then destroyed. In 1985 there were 656 reported insurgent incidents in South West Africa and in 1986 there were 476 with 644 PLAN operatives killed by security forces. In January 1987 SWAPO insurgents became active in the white farming areas for the first time in four years. A massive Soviet led Angolan army offensive against UNITA in southeastern Angola in 1987 was defeated by a South African mechanized intervention called Operation Modular. However, the subsequent South African push toward the Cuban and Angolan defensive positions around Cuito Cuanavale in late 1987 and early 1988 stalled and the movement of Cuban/Angolan forces in the southwest toward the border with South West Africa facilitated the success of American sponsored negotiations. In a staged plan supervised by the UN, South African forces withdrew from southern Angola and all of South West Africa in 1989, the Cubans pulled out of southern Angola around the same time and left the country in 1991, and Namibia became independent in 1990. The last combat in South West Africa occurred in April 1989 when around 1500 PLAN fighters crossed from Angola to establish a military presence before the upcoming elections and were intercepted by South African forces authorized by the UN to take action. Nevertheless, SWAPO won the election and became the first government of independent Namibia.[3]

The history and working culture of South African security forces helped shape their use of counter-insurgency tracking in South West Africa. In 1912, the two year old Union of South Africa created embryonic national security institutions in the form of the South African Police (SAP) and Union Defence Force (UDF). The SAP had a strongly para-military history as it had originated from nineteenth century formations like the Cape Colony's Frontier Armed and Mounted Police (FAMP) that had been involved in the Cape-Xhosa Wars and the South African Constabulary that had fought against Boer guerrillas in the Transvaal and Orange Free State during the Second Anglo-Boer War of 1899–1902. While other British settler dominions such as Canada and Australia shared this history of conquest era para-military policing, the SAP continued this tradition throughout the twentieth century. During the Second World War two all-white SAP battalions fought as front-line infantry in the North Africa campaign, draconian apartheid laws of the 1950s expanded police powers to enable them to suppress black protest and ultimately insurgency, routine law enforcement became a secondary duty, an SAP para-military contingent was dispatched to assist Rhodesian security forces in the late 1960s and military-style counter-insurgency training was introduced at the start of the 1970s. Therefore, when command of counter-insurgency efforts in South West Africa was transferred to the South African military in 1974, the politically influential SAP refused to completely withdraw from the arena and continued to field its own para-military counter-insurgency units. It is also important to note that the SAP had a long history of employing black constables and detectives, central to supervising black communities that stretched back to its nineteenth century forebearers.[4] In its early years the UDF was characterized by a combination of British imperial and Boer republican military traditions, and consisted of a small permanent component of South African Mounted Rifles, a part-time Citizen Force of British style regiments based mostly in towns and cities, and part-time rural and mostly Afrikaner commandos. Since military service was linked to citizenship rights, the UDF was an all-white institution and unarmed black participation was subject to parliamentary approval. However, the need for military manpower during the world wars led to some adjustments to this policy. A shortage of white volunteers for the East Africa campaign of the First World War prompted the creation of an armed component of the Cape Corps made up of mixed race or 'Coloured' men. Furthermore, during both world wars mixed race and black South African men were formed into large units of uniformed but unarmed military labor which served in various overseas theatres. The experience of fighting alongside British forces during the world wars greatly strengthened the British element of UDF military culture. However, the election of the Afrikaner Nationalist Party in 1948, which was strongly republican and implemented a thorough form of racial segregation called apartheid, caused the UDF to distance itself from British traditions, and mixed race and black military service was completely banned. In 1957, predicting the declaration of a republic four years later, the UDF was renamed SADF. With the rise of African nationalist protest and insurgency in the 1960s, the SADF experienced major expansion and institutional change. During the 1950s a limited number of white young men were selected by ballot of short periods of compulsory military training. In the 1960s this system was expanded and by 1968 all white young men had to complete one year of military service followed by yearly call-ups. Between 1960 and 1964 the SADF permanent force grew from 9000 to 15 000 and the number of national servicemen increased from 2000 to 20 000. Inspired by the counter-insurgency campaigns of the French in Algeria and the Americans in Vietnam which utilized indigenous troops who knew local languages and geography, as well as continuing military manpower problems caused by the relatively small size of the white population, the SADF began to rethink its white only policy. In 1963 the Cape Corps was re-established with an infantry battalion formed in the early 1970s, in 1973 the Army began to train black personnel who were quickly formed into the first of a number of black infantry battalions and in 1975 the Navy opened its doors to Indians. These troops became part of the volunteer permanent force, many advanced in rank and within a few years some Coloureds and Indians, and a very few blacks were commissioned as officers. Furthermore, the South African policy of granting independence to a series of small and impoverished black ethnic homelands, accelerated after the nationwide uprising by black youth in 1976, increased black military service and career opportunities within the context of the defence forces of these Bantustans.

During the 1960s and 1970s South Africa bought advanced military aircraft, launched a local armaments industry to counter mounting international sanctions, conducted its first large scale military exercises, and established Special Forces and conventional parachute units. As South African military involvement in South West Africa and Angola escalated, the period of national service for white men increased to two years in 1978 which meant that the number of active service SADF personnel mushroomed from 50 000 in 1970 to 150 000 in 1980 to 200 000 in 1985. All of this caused South African military spending to rise from R257 million in 1970–1 to R2.4 billion in 1980–1 to R4.8 billion in 1985–6.[5] By the middle of the 1970s, around the time it took control of counter-insurgency operations in South West Africa, the SADF had become large and sophisticated, and was transforming its hitherto narrow and racist military culture into something more pragmatic though still white dominated. These institutional changes came at just the right time to facilitate the development of SADF combat tracking in South West Africa.

Early Tracking Forces and Operations

In South West Africa there was a long history of white police, hunters and farmers employing trackers from the marginalized Bushman minority. As discussed in chapter two, police in South West Africa had engaged civilian Bushmen trackers since the 1920s. In the 1950s white farmers used 'tame' Bushmen trackers to locate other Bushmen communities where people were abducted as forced labour.[6] When SWAPO infiltration began in the late 1960s the SAP was in charge of counter-insurgency in the territory and it appears that in 1966 or 1967 it used Bushmen trackers and tracker dogs to pursue insurgents.[7] Since the very early twentieth century white police in South Africa had employed tracker dogs to pursue and even attempt to identify black stock thieves in rural areas with handlers and animals, including some from other areas of the British Empire such as insurgency plagued Palestine during the 1930s, trained at a formal dog school in the Transvaal.[8] At times police units on the trail of SWAPO during the late 1960s and very early 1970s also engaged trackers from the National Parks service as much of the Caprivi Strip through which the guerrillas usually passed had recently been declared a game reserve.[9] Around 1970 the SAP began to form small units of trackers from among some of the most isolated and least politicized Bushman groups in northern South West Africa. Based at a series of camps spaced about 40 to 60 kilometres apart, these trackers conducted daily patrols of the frontier and reported signs of insurgent infiltration. The SAP also dispatched Bushmen trackers into neighbouring Botswana to surreptitiously gather information on insurgents and recruit new members.[10] The police in Botswana also employed civilian Bushmen trackers, perhaps some of the same ones, to monitor liberation movements. For example, when an armed South African African ANC reconnaissance detachment was smuggled across the Zambian border at Kazungula in September 1966, just prior to Botswana's independence, the Bechuanaland police assigned Bushmen trackers to keep an eye on them until they were arrested by game rangers.[11] Derek Franklin, a British veteran of the war against Mau Mau in the 1950s and head of the Botswana Police Special Branch in the late 1970s, maintained 'that the South Africans were actually employing Botswana Bushmen, providing them with cash and clothing in an arid region where job opportunities were almost nil. The enterprising South Africans were hiring the Bushmen on contract as trackers to hunt down the elusive SWAPO gangs and in the process they armed their recruits with automatic rifles'.[12]

Based on its experience in Rhodesia's Zambezi Valley and unfolding events in South West Africa, the SAP established a para-military counter-insurgency school at Maleoskop in South Africa in 1970 which provided training on infantry weapons and infantry skills such as patrolling, ambushing and counter-ambushing, attacking an enemy base, helicopter operations, coordination of air support and follow-up operations which would have included tracking. All new police constables were required to attend this course and SAP members were sent there on refresher courses before their regular tours of duty in the operational area of northern South West Africa. Furthermore, an elite Special Task Force with advanced training in urban and rural operations was created in the mid-1970s. All this meant that during the 1970s the SAP developed tracking training programs in South Africa. By the early 1980s the SAP ran a six week basic tracking and survival course near Potgietersrus, also the site of a police dog school, and another six week advanced course in the Kruger National Park that culminated in a candidate performing a 250 kilometre long bush walk with only a rifle. Tracking instructors included a former member of the Rhodesian Army, and Bushmen and Ovambos from South West Africa.[13]

In 1972 and 1973 PLAN accelerated its activities in northern South West Africa. Around this time the SAP formed 'Cobra Teams' each consisting of five white personnel and one black special constable/interpreter which were transported by helicopter into northern South West Africa for week-long patrols to collect intelligence on SWAPO that was reported to the SADF for reaction. The Cobra Teams often worked with local unpaid special constables, called 'Oscar Zulus' or OZs, who were armed with obsolete bolt-action rifles, wore civilian cloths and patrolled the cleared border with Angola. Among the OZs were many Bushmen trackers. In December 1974 police Sergeant Chris Nel conducted the first formal three month training course at Ohangwena, near the Angolan border, for 60 Bushmen OZs in which they were taught military aspects of reconnaissance and tracking, and the handling of newly issued semi-automatic rifles. The second training class consisted of 60 Ovambo OZs who were issued uniforms and rations, and considered by police instructors as superior trackers to the Bushmen. Now paid, these trained special constables were loaned to police and army units as trackers and interpreters.[14] Although the Ovambo people were not historically hunter-gatherers but lived in settled agricultural and pastoral communities, Ovambo children learned to track while tending their parents' cattle as they had to identify specific animals by their hoof prints and they played tracking games by following small wild animals and analyzing people's footprints.[15]

Given SWAPO's transit through southern Angola, the SADF cooperated with the Portuguese military across the border. In the late 1960s and early 1970s, South African military support for the Portuguese in Angola included provision of medical assistance, intelligence, helicopter transport and 'a small number of experienced SADF trackers' dressed in Portuguese army uniforms who hunted UNITA insurgents as they were assisting SWAPO.[16] During the early 1970s South African Special Forces, in conjunction with Portuguese forces and local Bushmen trackers, patrolled southeastern Angola in search of infiltration routes and water holes used by SWAPO insurgents who were moving through the area from western Zambia on their way to Ovamboland in northern South West Africa.[17] In April 1974, with a military coup in Portugal that would mean the independence of Angola and therefore the expansion of the insurgency south of the border, the SADF took command of counter-insurgency operations in South West Africa. In September 1974, the SADF established a camp in South West Africa's Caprivi Strip for Angolan Bushmen who had been 'flechas' (arrows or trackers – see chapter 4) in the departing Portuguese security forces and were crossing the border to seek sanctuary from their nationalist enemies. The Bushmen were formed into an ad hoc military unit that patrolled into southeastern Angola to hunt SWAPO and some were attached to SADF units as trackers.[18] During Operation Savannah in 1975 and early 1976, the SADF intervention in Angola was enacted by two battle groups of Angolan troops; Battle Group Alpha consisted of the Angolan Bushmen unit based in Caprivi and Battle Group Bravo comprised Bantu-speaking Angolans who had been members of FNLA and UNITA. Alpha functioned as a motorized infantry force and recruited additional Bushmen including more former flechas who feared reprisals, others seeking employment and still others who were enlisted at gunpoint. After the failure of Operation Savannah, battle groups Alpha and Bravo were based in northern South West Africa and became the SADF's first primarily black but still white led fighting units designated 31 and 32 Battalion, respectively. These battalions would serve as models for the expansion of black combat forces which the SADF was beginning to see as important in fighting a counter-insurgency war in southern Africa. In the late 1970s, the Caprivi based 31 Battalion was a conventional infantry unit the members of which had a reputation as skilled trackers given the long-standing image of the Bushmen, and an aggressive and sometimes coercive recruiting campaign was launched to fill its ranks from within South West Africa.[19] It has been suggested that the main military advantage possessed by the Bushmen was not their tracking ability but their pariah status in wider South West African society which enabled the SADF to mobilize them against insurgents from larger groups, particularly the Ovambo.[20] Social change imposed by the colonial state such as through re-settlement schemes meant that most Bushmen born after around 1960 did not possess the same refined tracking or bush survival skills as their forbearers.[21] However, it is clear from many accounts that some Bushmen recruited into the South African security forces were outstanding trackers.

With the advent of the MPLA regime in Luanda and the establishment of SWAPO bases in southern Angola, there was a dramatic increase in insurgent activity in northern South West Africa during the late 1970s and their tactics changed from 'hit-and-run' attacks in the border zone to deeper infiltrations meant to attack white farms further south.[22] While the South African security forces responded by increasing the number of personnel and resources in the operational area, the guerrillas' elusiveness and their retreat into Angola when detected led to a number of initiatives related to tracking. In 1976 the South Africans, to detect insurgent infiltration, removed around 50 000 people from South West Africa's frontier with Angola and created a one kilometre wide depopulated strip called the 'Yati'.[23] Between the border and the white farming areas to the south a network of sandy roads called 'cutlines' or 'kaplyne' were regularly swept by military vehicles dragging trees and routinely patrolled for footprints.[24] The central feature of the SADF counter-insurgency campaign was a continuous series of patrols meant to dominate the operational area and collect information. Patrols ranged in size from ten men to a company, lasted from several hours or days to six or eight weeks, and were conducted on foot or by men mounted on horses, motorcycles or larger vehicles. From the late 1970s, as discussed below, patrols that discovered evidence of PLAN infiltration, frequently tracks, would call in a helicopter borne or vehicle mounted reaction force to take over the pursuit. This determination to locate insurgents led to 'the development of tracking as a fine art in the SADF'.[25] PLAN field commander Johannes Gaomab explained that 'Soon the SADF was on our track. The problem with guerrilla war is that you have to walk, and if you walk, you leave spoor. The SADF had trackers patrolling up and down the border with Angola'.[26] Although the sandy soil of northern South West Africa made insurgent anti-tracking difficult, the flat and featureless terrain enhanced mobility on foot and bush often inhibited long-distance visibility. Furthermore, PLAN's predominantly Ovambo fighters blended with the rural population of Ovamboland.[27] Since South African sponsored UNITA was based in southeastern Angola, SWAPO did not maintain staging areas there which meant that it was difficult for its guerrillas to penetrate the adjacent Kavango area of northeastern South West Africa. Insurgents intending to enter Kavango had to cross the border into north-central Ovamboland and then move east across the dry open ground of western Kavango during which time they were vulnerable to observation making for the bush cover of eastern Kavango.[28]

From the first arrival of former flechas in South West Africa, SADF authorities saw the potential of utilizing the legendary Bushmen tracking abilities in the counter-insurgency campaign. In 1974 Major General Fritz Loots, Chief of SADF Special Operations sent junior officers Tom van Deventer and Pinkie Coetzee to train the Bushmen soldiers of newly created Battle Group Alpha as military trackers but this was delayed by the launch of Operation Savannah. Immediately after Savannah, the SADF attempted to organize tracker training by sending an infantry school instructor to learn tracking at the Natal Parks Board, a tracking conference was held to prepare a handbook and basic courses were run at a camp on the Cuando (Kwando) River in the Caprivi Strip. However, the Bushmen soldiers from 31 Battalion sent on these courses learned very little as the level of instruction was low. The growth of PLAN infiltration of northern South West Africa in 1976 prompted the attachment of 31 Battalion Bushmen trackers under van Deventer to the SAP and during Operation Cobra, an SADF attempt in May 1976 to find and destroy the PLAN presence in Ovamboland and southern Angola, they were first employed by other SADF units. In one incident during Cobra, a Bushman soldier tracked an insurgent group from a helicopter but the practice was not continued as the low and slow flying aircraft was susceptible to ground fire. After Cobra, select Bushmen were sent to Ovamboland to assist SADF units counter SWAPO infiltration which led to the 1977 creation of a separate SADF Tracker Unit made up of some of the best trackers taken from 31 Battalion and commanded by van Deventer.[29] After specialized training by SADF Special Forces, 40 Bushmen from 31 Battalion formed the unit's Reconnaissance or Recce Wing in February 1977. Divided into four teams each led by a white officer, the Battalion's Recce Wing was heavily used to patrol southern Angola, occasionally accompanied Special Forces teams and killed 133 SWAPO insurgents between 1977 and 1981. By the 1980s the battalion regularly offered a four week course that taught 50 to 120 Bushmen soldiers the modern tactical application of tracking including impressions made in the ground by weapons, insurgent anti-tracking techniques and the risks of landmines. Graduates of this course tracked for other security force units and some went for more advanced survival training elsewhere. In 1978 a few of the Recce Wing's best Bushmen trackers were selected for parachute training, incorporated into Special Forces and employed in covert operations in Rhodesia. By 1983 the 31 Battalion Recce Wing had expanded to six teams each with two whites and four Bushmen and consisted of 'the most expert trackers'.

White leadership was considered essential as the presence of two whites in a team allowed it two divide into two independent groups. Unit historian Ian Uys claims that 'They complemented each other as the white could often interpret the signs the Bushmen trackers found'.[30] Referring to the Recce Wing, anthropologists Gordon and Douglas maintain 'it was these soldiers and trackers who were regularly deployed with other battalions and upon whom the martial reputation of the battalion ultimately depended'.[31]

The presence of Bushmen trackers not only improved the chances that SADF units would locate insurgents but sought to mobilize popular myths about their superior senses as a psychological weapon to instil fear among the enemy and inspire confidence in young white national servicemen working in an alien environment.[32] Accounts by PLAN veterans, well aware of stereotypes about the Bushmen, describe the effectiveness of Bushmen trackers working for the South African security forces. According to Oswin Namakalu, 'The SADF recruited the Khoisan as scouts and trackers of guerrilla movements. The San, who are traditionally hunters, are good at tracking'.[33] Peter Ekandjo, who spent over five years as a guerrilla reconnaissance specialist in northern South West Africa in the late 1970s and early 1980s, remembers that 'Bushmen gave us a tough time, they were good trackers'.[34] South African national servicemen told journalists that patrols accompanied by ever alert Bushmen trackers did not need to use maps and compasses for navigation or post sentries at night, and that the presence of these auxiliaries meant their chance of dying was much less. Humorous stories about the Bushmen also served to raise morale as it was claimed that they sometimes abandoned the trail of insurgents to pursue a source of wild honey or broke noise discipline by uncontrollably laughing at the sight of a hyena.[35] A 1978 issue of the SADF magazine Paratus published a drawing of a white South African soldier being directed by a pointing black tracker with a caption reading 'Bushman Tracker: These small unobtrusive little fellows are worth their weight in gold'.[36] As discussed in chapter one, the nineteenth century tradition of comparing Bushmen to dogs continued during the Border War. Paratus boasted that 'Once an insurgent was trailed for five days while he employed every trick in the book to shake off the human bloodhound on his track – but to no avail'.[37] A right-wing American journalist wrote that 'If you've never seen a two-legged bloodhound at work, come to South West Africa and watch the Bushman. Actually, the Bushman puts the bloodhound to shame'.[38]

Numerous accounts by white South African veterans of the Border War mention how impressed they were by indigenous trackers, particularly Bushmen. An anonymous white citizen force soldier with the Transvaal Scottish, a citizen force unit, called up in 1976 described his experience with Bushmen trackers in the Caprivi Strip:

Occasionally, the Bushmen assigned to patrols would pick up tracks, and we would gear up for an attack or a rapid response. Usually these were just probes, and the tracks would fast disappear back across the border. This was the general scheme of things in '76 in the Caprivi. It was pretty quiet, and thinking back we were not really in a high state of "mental" readiness ... Most of the patrols were just a rifle platoon, guided by a Bushman, with perhaps a 60mm (mortar) section as additional support. They would zig zag across their assigned patrol area from drop off to pick up points trying to find spoor. One group had a K9 corps handler and dog assigned on one occasion ... The Bushmen trackers intrigued me, they were tough little guys, who seemed to have a sixth and seventh sense.[39]

SADF medical officer Anthony Feinstein, in 1983, observed that a police unit based at Tsandi in Ovamboland retained the services of around 20 African trackers referred to as 'buddies' who were armed but only partly trained. Recalling a mechanized patrol pushing its way through the bush, Feinstein wrote:

> At intervals, the patrol paused and the buddies would alight to look for tracks. I was intrigued by their ability in this regard. A few of them would huddle over a spot of earth that was meaningless to my eyes, confer briefly, reach unanimity quickly, call out their instructions in a combination of broken English and Afrikaans liberally interspersed with obscenities and the patrol would move on. It was soon apparent that when on patrol, we were totally reliant on them for information and directions. Such was our dependence, we would never have made it back to camp in the evenings without their guidance.[40]

In 1984 South African infantryman Tim Ramsden was part of a routine mechanized patrol around the Etosha National Park area which was accompanied by two diminutive Bushmen trackers from 201 Battalion (a new designation for 31 Battalion as part of the SWATF):

> The San are amazing trackers and could read spoor from Buffels (vehicles) travelling at 30 kph. Our eyes darted from the road and back to our Bushman, who scanned the ground with deep concentration. Suddenly he signalled a sighting and Bennie immediately stopped the vehicle. We clambered out, following the tracker's lead as he stooped into a squatting position and examined a boot print. Many telltale signs indicated its age. Insects had crossed the print, leaves had fallen into it and the wind had distorted the outline of the indentation. Reading the information, our Bushman tracker told us that our enemy had passed a couple of days ago.[41]

Bushmen trackers were also employed by part-time white commandos. In 1978, with the first insurgent attacks in the white farming area around Tsumeb and Grootfontein, the local all-white commando began training and arming black farm workers who were incorporated into local defence plans. At the same time, the commando founded its own tracking unit, including local Bushmen who were given military training, to assist security forces in the area. In 1981, the same year the commando was renamed the Etosha Area Defence Unit, some of its Bushmen trackers led by a white farmer who was also their civilian employer first pursued insurgents. These Bushmen were paid but some felt that if they refused military demands they would be accused of supporting the insurgents.[42] Although south of the main operational area in Ovamboland, combat tracking in this area could be dangerous as an anonymous white farmer and 'experienced tracker' discovered in the early 1980s:

> I was following a spoor with my trackers and had repeatedly warned them to stay back while I walked in front. We were hot on the trail of a band of terrs and I stepped on a well-camouflaged cable of a "Black Widow" anti-personnel mine. Fortunately, I was standing close to the explosion so my legs took the force of the blast. One of my trackers, who was standing near me at the time, was also injured and he too went down.[43]

With the rise of African nationalist insurgency in the Southern Africa region during the 1960s and 1970s, the SADF began to utilize the superior senses and physical abilities of animals for military purposes. In 1964 it opened a dog school, initially at Voortrekkerhoogte near Pretoria but eventually moved to Bourke's Luck in the eastern Transvaal, to train dogs and handlers for security work, explosive detection and scent tracking.[44] During the 1980s the SADF experimented with developing a special breed of 'wolf-dog' that would more effectively pursue insurgents but the pads of its feet proved too soft for conditions in South West Africa and the use of small boots was embarrassing.[45] In 1974 the SADF established an Equestrian Centre in Potchefstroom to breed and train horses, and train white national servicemen as mounted infantry for counter-insurgency operations. The founding instructor at the centre was Major Peter Stark who had been taught to track as a young man in South West Africa by a Haikom Bushman named Willie who he described as his 'black father'. Stark also had been formally trained in equestrian sports in West Germany and had been head warden at Etosha National Park and chief game warden of South West Africa. He was introduced to senior SADF officers while running tracking and bush survival courses for the military in the early 1970s. An ardent tracker, Stark wrote that 'I continually practised my tracking skills, and later, when I worked for Nature Conservation, I often outsmarted the Bushmen'.[46] Some army officers posted to the Equestrian Centre also gained expertise by visiting the Grey's Scouts in Rhodesia and the Army Cavalry School in Chile. The South Africans were certainly aware that the Portuguese army had used horse mounted troops to patrol the vast interior of Angola but by this time they had withdrawn from Africa. The military use of horses appealed to the Afrikaner memory of mounted commandos during the Second Anglo-Boer War and the riding interests of senior politicians and SADF officers. During counter-insurgency in South West Africa, the elevated height of horse mounted infantry helped them spot enemy spoor during reconnaissance patrols, horses could provide early warning of enemy ambush, horses' speed helped security forces catch up with insurgents on foot and horses' relative silence conferred an element of surprise. Mounted infantry platoons were assigned to infantry battalions and bases across South West Africa. Mounted patrols were responsible for constant monitoring of specific two kilometre long sections of the various cut-lines and were accompanied by African trackers, including Bushmen, who were trained to ride. Since horses and riders made good targets and were vulnerable to ambush, a mounted patrol on the trail of insurgents would call in a mechanized unit to take over the pursuit when it was believed they were getting close to the enemy. Frequently, mounted detachments operated 500 to 1000 meters in advance of larger mechanized forces.

In 1977 the SADF consolidated several reaction units into 101 Specialist Unit, based at Oshivelo near the northeastern border of Etosha National Park, which combined human trackers, and scent tracker and mine detector dogs with the mobility of horse and motorcycle mounted infantry. This unit became part of the SWATF, established in 1980 to localize command of the counter-insurgency effort, and was renamed 1 South West African Specialist Unit (SWASPES) with a permanent base built at Otavi just south of Etosha in the white farming area. The unit was divided into three elements; horse mounted infantry, motorcycle mounted infantry and trackers which included human visual trackers and tracking dogs with their handlers. While the horse mounted infantry were trained at the Equestrian Centre and the dogs and dog handlers at the Dog School, the motorcyclists and trackers were trained locally by SWASPES. Since the tracking element was the largest of the three, most SWASPES personnel undertook tracker training at a Spartan bush camp located six kilometres away from the unit's base. The basic tracking course emphasized physical fitness and determination, taught rudimentary spoor identification, enemy anti-tracking techniques and bush survival, involved 20 kilometres of practical tracking per day and finished with a test in which candidates had to follow a difficult spoor for one kilometre. The advanced course taught anti-tracking, back-tracking, team tracking, environmental knowledge and bush survival, and culminated in a practical survival test. Members of other units took these courses including Bushmen from 201 Battalion who did the advanced segment and then served as assistant instructors. SWASPES unit symbols emphasized the tracker mission as those who graduated from tracker courses wore a honey badger – an animal famous for finding bees' nests – patch on their uniform and the unit crest featured the print of a bare human foot. In the best circumstances, it was possible for a team of human trackers to find insurgent spoor which horse mounted infantry would track during the day and close the distance with the enemy so that tracker dogs could pick up the scent and continue pursuit during the night when visual tracking does not work. In the late 1980s the unit began experimenting with using packs of Irish wolfhounds, capable of running 15 kilometres an hour for four hours, to led horse mounted infantry toward insurgents. Attached to operational battalions, elements of SWASPES were employed in the mountainous parts of

Kaokoland in the northwest where larger vehicles were of limited utility and the level of PLAN infiltration fairly low. Although there were some problems with South African commanders not knowing how to employ tracker dogs including failing to realize that dogs tracked by smell, it appears that at least 60 percent of the unit's operations resulted in insurgent contact. However, PLAN claimed to have killed or captured many SWASPES horses, and put some to work pulling wagons.[47] As in Rhodesia, the use of SADF tracker dogs was also problematic. According PLAN veteran Ekandjo, South African security force dogs were not used in the thick bush of eastern Ovamboland after around 1979 as many of the expensive animals were killed in ambushes or by landmines. He maintains that horses, dogs and motorcycles were more commonly used in semi-open western Ovamboland, and that PLAN fighters sometimes sprinkled pepper on the ground to confuse tracker dogs.[48] A 1985 incident illustrates some of the difficulties with tracking dogs. Two army tracker dogs were flown by helicopter to the scene of a police pursuit of a small unit of PLAN insurgents several of whom had been wounded. After disturbing the insurgents' spoor which frustrated the police trackers, the dogs ran into the bush where one was killed with a bayonet by a concealed guerrilla who then escaped with several others.[49]

The Tracking War

Given SWAPO attacks on Ovambo chiefs and government supporters during the 1970s, the police and army cooperated to transform the OZ program into an Ovambo Home Guard which, by the end of 1978, numbered 3000 trained special constables many of whom were seconded to regular units. In late 1978 Ovambo Home Guard training was shifted to a more developed camp at Tsandi, training capacity was expanded to seven platoons of 40–45 men each and candidates arrived from neighbouring Kaokoland and Kavango. In the middle of 1978, with increased SWAPO infiltration from Angola, South African security chiefs decided to establish a special unit modelled on the Rhodesian Selous Scouts to conduct pseudo-operations with the police gathering intelligence and the SADF's Special Forces (Recce Commandos) organizing pseudo teams. In January 1979 the SAP established a special unit called Operation K or Koevoet, Afrikaans for 'crowbar', at Oshakati in Ovamboland as its contribution to the project. Under Colonel 'Sterk' (Strong) Hans Dreyer, the embryonic Koevoet consisted of six white officers including a young Lieutenant Eugene de Kock, who had fought in Rhodesia and later became the apartheid state's most notorious assassin, and 60 of the best Ovambo Home Guard special constables who were skilled trackers. Studying the existing counter-insurgency approach, Koevoet officers observed that regular patrols by mostly white police and entirely white national servicemen from South Africa were easily avoided by SWAPO, reports about insurgents were not followed up, and police and military personnel failed to recognize insurgent tracks and rarely employed local trackers and interpreters. The original pseudo-team approach did not materialize as the Recces were too involved with external operations to capture insurgents for the police to 'turn', police and army culture were incompatible, and SWAPO groups were too well established within the country to be effectively infiltrated by imposters. In June 1979 Koevoet Sergeant Chris Nel and 30 special constables were sent to a white farm at Tsintsabis that had been attacked by PLAN and, without vehicles, tracked a group of 30 insurgents for about 200 kilometres over five days that led to an engagement in which one Koevoet member and eight guerrillas were killed. During the operation another 60 strong PLAN unit tracked Nel's team but was driven into Angola by de Kock and 30 Koevoet special constables landed by helicopter.[50] Years later Dreyer told an American journalist that 'Everyone was amazed that we could follow spoor for that long ... I knew then I had something golden. It wasn't long before we were killing 50 to 80 terrs (insurgents) a month'.[51]

Inspired by the success of the Tsintsabis operation and his own experiences in Mozambique and Rhodesia, Dreyer instructed de Kock to form two Koevoet fighting groups from among some former FNLA fighters recruited from 32 Battalion and the original 60 special constables who received a month's combat training by Special Forces in Caprivi. With more Ovambo Home Guard recruits and a new training base at Ongwediva, Koevoet quickly expanded to four fighting groups with a total of eight white policemen and 304 black special constables. Rather than patrolling set areas, the teams deployed and tracked insurgents based on specific intelligence often obtained by Koevoet's investigative section. The early Koevoet tracking operations in eastern Ovamboland in late 1979 were conducted on foot with a few Hippo mine-protected vehicles used to transport teams from a temporary base in the bush to the scene of insurgent spoor. It was at this time that group leader Sergeant Chris de Wit accelerated pursuit by driving behind trackers with a Hippo loaded with water and additional trackers who would relieve those on the ground.[52] In February 1980 de Kock, taking advantage of the semi-open terrain, began practising with three or four lightly equipped Ovambo special constables tracking while running alongside a moving vehicle in a mechanized version of persistence hunting.[53] Although it has never been recognized, Koevoet leaders may have been inspired by a similar tracking technique used by game rangers. Some years earlier, in Etosha National Park, head warden Peter Stark and his Bushmen trackers assisted the police to pursue a group of escaped convicts. Stark wrote that 'Two trackers were always on the ground, and the rest rode along on the back of my vehicle. When the two on the ground tired, they scrambled onto the load-bed, and rested men took over the pursuit'.[54] By the end of 1981 Koevoet had combined its investigative and fighting teams to form 20 teams each consisting of 30 men and four new Cas-

spir mine-protected infantry fighting vehicles and a logistics vehicle. Divided
into two groups, half the teams would spend a week in the bush and were then
relieved by the other half fresh from rest and maintenance.[55] The teams enjoyed
considerable autonomy and were encouraged to experiment. Collecting information from informants and looking for spoor as they travelled, Koevoet teams
would circle a village at more than a kilometre to avoid masses of civilian footprints and determine if insurgents had visited recently. Koevoet developed a
standard procedure of having some trackers on foot followed closely by others
in vehicles, leapfrogging by dispatching two vehicles sometimes up to ten kilometres forward to find the continuation of a line of spoor, speculatively firing
mortars and grenade launchers in the suspected direction of enemy to panic
them into running which would leave more obvious tracks and discarded equipment, and upon closing with insurgents they called for helicopter gunships or
a spotter plane for support (see below).[56] Moreover, the unit began employing
captured SWAPO insurgents who changed sides bringing with them knowledge
about their former colleagues' methods. Indeed, Koevoet went out of its way
to apprehend particularly skilled anti-trackers such as 'Jack' who, around 1983,
was pursued for four days and nights from Ovamboland into southern Angola
where he was captured.[57] Since Koevoet mostly operated in Ovamboland which
was the centre of insurgent activity and recruiting, unit membership was dominated by Ovambo trackers who were familiar with the area and its people. Very
few Bushmen worked within Koevoet where both whites and Ovambos stereotyped them as inferior trackers and cowards.[58] Koevoet was also flexible and
when appropriate, such as when searching a village or when silence was needed,
a team would deploy most of its trackers on the ground to locate insurgent spoor.
They also set night ambushes on suspected enemy routes and sometimes African
trackers posed as PLAN insurgents to collect information among communities. The combination of trackers on the ground following spoor and vehicle
commanders manning heavy weapons from a high vantage point was also effective. Once Koevoet began routinely pursuing insurgents across the border into
Angola in the mid-1980s, the increased chance of ambush meant that a single
vehicle would not leapfrog very far but rather several heavily armed Casspirs
loaded with trackers were sent forward. Poorly paid trackers were given additional motivation by cash bonuses for dead guerrillas and captured weapons.
During 1986 Koevoet teams adjusted their tactics by deploying fewer trackers
on the ground, 12 when dismounted and five when mounted, as more personnel tended to walk on spoor and cause distractions. This also reduced potential
casualties from ambush or landmine.[59] Not all Koevoet experiments worked as
attempts to use tracking dogs proved useless given the area's extreme heat which
became much worse inside a vehicle, noise from the vehicles and stress of combat. One Koevoet team leader went so far as to make small boots to protect his

tracker dog's feet from the hot metal floor of the Casspir but the ineffectual animal was eventually sent back to South Africa. Environment determined Koevoet tactics. In western Ovamboland, a semi-open area with hard ground difficult for tracking and where people were supportive of SWAPO, Koevoet teams spread out and hoped to encounter insurgents or their spoor by accident. In eastern Ovamboland, with its thick bush and less politicized inhabitants, they patrolled and gathered information from more cooperative locals.[60] Since vehicles could not operate in much of mountainous Kaokoland, Koevoet teams working there employed infantry tactics such as observation posts and night ambushes, and dismounted tracking teams were sometimes leapfrogged ahead by air force helicopters. In sparsely populated Kavango, Koevoet often shelled the relatively large insurgent groups operating there with mortars to prompt them to split up so that smaller groups or individuals could be tracked with less danger of ambush.[61] Jack Greeff, a highly decorated South African Special Forces operator who worked with Koevoet, explained that 'the reason for their success was their tracking ability. This was something that most SADF units did not have ... Tracking, tracking and more tracking was their motto'.[62] By the end of the conflict in 1989, Koevoet had 3000 personnel and 42 combat teams, had fought 1600 engagements in which 3200 alleged insurgents were killed at a cost of 161 of its own men killed and 950 wounded.[63] Of course, such lethal success attracted attention and Koevoet became accused of brutalizing civilians at around the same time that the SADF counter-insurgency effort abandoned attempts to win the 'hearts and minds' of local people.[64] The top SADF leadership greatly resented police involvement in the campaign and often complained about Koevoet's operational autonomy and aggressive approach. According to former SADF chief General Constand Viljoen, Koevoet 'had a cruelty about them that certainly didn't further the hearts and minds of the people ... in a revolutionary war it is not a case of how many people you kill but rather the battle for the minds of the people'.[65]

Despite SADF commanders' distaste for Koevoet's methods, they eagerly copied them during the early 1980s. This happened at the same time that large conventional SADF cross-border incursions and support for UNITA rebels forced PLAN to withdraw its bases deeper into Angola which restricted infiltration of South West Africa. Several SADF battalions specialized in fielding mechanized 'Romeo Mike' (RM) teams named after the Afrikaans term 'Reaksie-mag' or 'Reaction Force'. According to former South African Air Force (SAAF) Brigadier Dick Lord, 'In an attempt to improve their results the army decided to form mobile Romeo Mike (RM) teams, based on the concept of the Koevoet Zulu teams'.[66] Beginning around 1980 201 'Bushman' Battalion (formerly 31 Battalion) organized six week rotations during which two companies remained at Omega Camp in Caprivi for rest and training while two others went into the bush where they broke into reaction teams that would be called to the scene of a security force encounter with PLAN to track fleeing insurgents. A 201 Battalion RM company consisted of seven teams of 20 men mounted in three Buffel mine-protected vehicles. Like Koevoet, a constant pursuit was maintained over a long distance by having trackers rotate from the ground to the vehicles, and trying to gain ground on the enemy by detachments speculatively driving ahead of the spoor. In late 1982 teams from 201 Battalion operating in the dense bush of eastern Ovamboland tracked a SWAPO Special Forces detachment for 278 kilometres, of which 190 kilometres were actually run on the ground, over three days and killed all seven of them in two engagements without any losses.[67]

In 1983 the SWATF's 101 Battalion was restructured with Koevoet organization, vehicles and tactics. This unit originated with the Ovambo Home Guard of 1974 some of whom were sent to the embryonic black 21 Battalion in South Africa for training in 1976 and formed into the ethnic Ovambo 35 Battalion in 1978. While 35 Battalion initially detached trackers and interpreters to other SADF units in northern South West Africa, it eventually fielded its own combat infantry companies and was designated 101 Battalion in 1981 as part of the new SWATF. By 1985 101 Battalion had two companies organized into Koevoet-style teams with Casspirs – it was the only army formation to use this hitherto police vehicle – and in a short time two more reaction companies were formed. By the end of the 1980s it consisted of four RM companies of 150 men each, two self-supporting dismounted infantry companies of 250 men each deployed for area protection, a horse and motorcycle mounted company, and an intelligence company of guides, interpreters and trackers permanently attached to SWATF Sector 10 (Kaokoland and Ovamboland) headquarters. The battalion conducted most of its own training, and the experienced Captain T. Ferreira developed standard procedures for mechanized RM operations and supervised regular refresher training for operational teams. It also boasted a large logistical component including dedicated vehicle maintenance detachments for each RM company. While almost all of the unit's officers and most support personnel were whites, some Ovambo troops quickly advanced to NCO status and several became junior officers in the context of reforms to the military racial hierarchy. This was a very large battalion which grew from around 1000 men in the mid-1980s to 2500 by 1989, 35% of its personnel were former SWAPO insurgents and soldiers' salaries meant that recruiting was never a problem in impoverished Ovamboland. Reinforcing its ethnic Ovambo identity which reflected apartheid divide-and-rule strategies, 101 Battalion built traditional beer huts and performed traditional songs during parades. Centred at Ondangwa and with several forward operating bases located very close to the northern border, the unit was extensively involved in southern Angola from 1986 engaging Angolan state and Cuban forces in conventional battles. By 1988 the battalion had accounted for 1400 enemy fatalities in South West Africa and Angola, and lost just 78 of its own men. An intense rivalry quickly developed between Koevoet and 101 Battalion the teams of which listened to each others' radio frequencies to get information on insurgent spoor and then tried to steal the other unit's prey. Since 101 Battalion consisted of Ovambo soldiers and white national servicemen, it had fewer skilled trackers than Koevoet and rarely leapfrogged very far forward. Unlike Koevoet teams led by experienced police sergeants, army RM teams were usually commanded by recently qualified young lieutenants who struggled to coordinate a mobile force and sometimes prematurely requested air support which alerted insurgents that they were being tracked.[68] Nevertheless, in 1986 two of the battalion's riflemen were awarded the Honoris Crux, South Africa's coveted military decoration for combat bravery, and it 'had the best combat record of all SWA and RSA units during the year'.[69] For white officers, Ovambo troops seemed to combine the super-senses of the hunter-gatherer Bushmen with the innate warrior characteristics of an imagined martial people like the Zulu. In 1985 Major J. Kruger, second in command of the battalion, told British reporters that his Ovambo soldiers 'come out of the field, they're natural trackers. They can follow a squirrel at the run, even if the enemy is busy anti-tracking and they are very aggressive and they are very well disciplined because of their tradition'.[70] The 101 Battalion trackers loaned to other units performed a dangerous job as during patrols or pursuits they were always in the lead and were often the first ones caught in an ambush or landmine explosion. South African paratrooper Granger Korff wrote about an incident in the early 1980s where his unit suddenly encountered five insurgents and 'The black 101 Battalion tracker walking point had spotted them first and had already opened up on them with his heavy 7.62 G3'.[71] Derek Kirkman, a national serviceman with 7 South African Infantry Battalion (7 SAI) in 1988 and 1989, remembered that 'We were then assigned a 101Bn tracker to help with patrols. He was pretty good. He really knew his stuff'.[72]

An anonymous medic attached to 101 Battalion in 1988 described how the unit operated:

> I had had some dealings with this unit during national service. They were made up of local Ovambo men. Most were ex terrorists that had for one reason or another changed side. We rode in Casspir vehicles and we went hunting. Most times we would be on the top of the vehicles rather than inside for a couple of reasons. Firstly the heat in the vehicles; above there was a nice wind from the movement of the K-Car (slang for Kill car). Also the guys had the theory that if you hit a boosted mine then the Casspir was stuffed. The guys inside were not going to be happy but the chaps on top would be thrown off by the blast.
>
> The vehicles normally had a .50 browning on the co-driver's window, either twin Brownings or a single 20mm on a top mount between and slightly back of the driver. SOP was that three or four cars (with 8 to 10 guys in each) would roam around or go to an area where there would be a report. The trackers would get down and hunt around or we would go to the local Kraal (a group of huts of a couple of families surrounded by a wooden stake wall) and "chat" to the headman.

These chats could be nice or nasty depending on the situation. I mean when one follows terr spoor (tracks) right into the Kraal and the headman then claims he never has heard of SWAPO let alone seen terrorists. So the chat then tended to be a little more stressful than normal. We were normally out for 2 weeks at a time, travelling from base to base to get fuel and rations as we needed.

When we got fresh spoor then the cars would split; three would stay on the spoor and one would run about 5 to 10km ahead in the general direction that the spoor was going as a backstop. One car could go left of the spoor (about 50 to 100m out) and the other right. The third car would travel up but slightly off the spoor in case of mines. The two outer cars would be about 100m in front of the middle car to act as protection walls for the tracker that would be running up the spoor by foot.

All of the guys on the car were expected to run the spoor in relays. One or two guys would track on foot usually with a pistol or R5. the others would sit in the car. After a time depending on the speed of the vehicles a guy would jump out of the back of the moving car – run up the spoor take the pistol from the front guy and carry on up the spoor. The other guy would slow down and the casspir would pass him and he would jump on the back.

Hunting was frightening and exciting. As I had a R5 the guys really liked me and borrowed it to run the spoor so I adopted the .50. I also ran the spoor (but not alone as to be honest I often could not see what these guys were following) as protection for the tracker. The camaraderie was great but it helped that I was a camper rather than NDP as they accepted me as I was older.[73]

South African soldiers described the work of tracking on the ground, accompanied by the lead vehicle, as 'running dog'.[74] A white national serviceman related his experience with the RM tracking system:

we went out the next day with Casspirs, ten Bushmen trackers in each, to pick up the spoor from the kaplyn (cut line) again. The terrs who had ambushed us had bomb-shelled (split up) completely; so we went back to the kaplyn looking for a larger group to track. We preferred to track a bunch of at least three or more individuals. We found spoor from a small group and then two of the trackers ran in front. Once they had the spoor they would keep on it, rotating trackers every hour or so, so that they were always fresh and fast. It was called hot pursuit. If we picked up tracks that were maybe 12 hours old we could catch up fast, and by the end of the day we would only be an hour or two behind. If we couldn't catch them that day, we camped for the night. Naturally the terrs didn't stop to sleep or eat but kept going. Of course, this meant that by morning were back to a 12-hour-old track. But the terrs knew if they crossed another kaplyn, we would not be far behind them. As they became more tired they starting dropping equipment, discarding whatever wasn't essential.

When we were only minutes behind, we called in the spotter plane. It could fly very slowly and very low, and it was fairly easy for the pilot to spot a person. When he spotted the guys we were chasing, he dropped a white phosphorous grenade as a marker. This rose and expanded and was easy to see over a fair distance. We put foot (acceler-ated). All the trackers back on board the Casspir; and it was a flat-out rush to reach that white cloud as fast as possible. We caught up with the terrs, four of them. The contact was brief. They were exhausted; they'd been running for days, so it was over quickly.[75]

The mechanized RM teams were not the only SADF reaction forces. At the end of the 1970s the SADF copied and adapted the Rhodesian Fire Force method of having an air mobile reaction team waiting at an airfield for rapid dispatch to the scene of an insurgent sighting. These teams were usually transported in three Puma helicopters accompanied by a helicopter gunship and could also include a small unit dropped by parachute from a Dakota transport aircraft. Refining the concept, South African Fire Forces conducted 'lunar operations' during moon lit nights by having a small team or paratroopers drop to intercept a suspicious civilian vehicle that was breaking the night-time curfew and had been observed by a helicopter. The paratroopers also sent out decoy vehicle convoys at night to attract an insurgent ambush which was then intercepted by a Fire Force already in the air. Once on the ground, Fire Force troops worked as normal dismounted infantry and disliked following insurgent spoor that was more than four to six hours old as they were vulnerable to ambush and landmines which could be prepared by insurgents if they had time. These Fire Force teams, copying another Rhodesian innovation, also used helicopters to position blocking groups along the suspected path of a fleeing enemy and leapfrog trackers ahead along a trail especially' in cases where an insurgent group began to split up or 'bombshell'.[76]

During the early 1980s Koevoet developed a standard air support tactic for tracking operations. Since Koevoet kept the SAAF informed of its actions, helicopters were prepositioned at a series of forward bases and were usually about 20 minutes flying time from any team in the field. A Koevoet team tracking insurgents updated the SAAF by radio as it gained ground on the enemy which allowed the pilots at the base to be briefed and then wait in their cockpits for dispatch to the scene. The ground team typically requested air support when it was about 20 or 30 minutes, which the SAAF believed amounted to about one kilometre, behind the enemy. Two helicopter gunships usually responded to a Koevoet team's call and remained at a distance until direct radio communication was established with the ground team which ignited a smoke grenade to indicate its location. The helicopters then flew two overlapping orbits over the team. The lead helicopter flew a narrow orbit at an altitude of around 200 feet and ranged about three kilometres ahead of the team to detect ambushes, and the other helicopter flew at between 600 and 800 feet and circled two to five kilometres from the team to discourage the insurgents from fleeing in different directions.[77] The helicopters flew at different altitudes to prevent mid-air collision and always used left-hand orbits as their weapons were mounted on the left door. They sometimes flew a 'Double-D' pattern with the straight part of the 'D' aligned with the direction of the enemy tracks. As one helicopter turned over the trackers, the other turned over the furthest point ahead of them which was how far the insurgents were estimated to be in the lead. Sometimes a helicopter picked up one of the ground team members to help with aerial observation. When the trackers observed distinctive signs of insurgents changing behaviour such as hiding under trees or running from tree to tree, they knew that the enemy was within the orbit of the helicopters. The movement of the lead helicopter was then changed to figure out how far ahead the insurgents were and therefore, how far ahead to send the advance vehicles to catch them. Accomplished Koevoet tracker Laurens Musore, who later applied some of these lessons while working as a security contractor in Iraq during the 2000s, explained that 'The use of helicopter gunships was a crucial point in tracking as it forced insurgents into hiding. When the gunship arrived the guy would walk normally and when it went away he would run and when it returned he would walk. That made the tracker suspect he was an insurgent'.[78] Once the ground team was very close to the enemy, the lead helicopter tightened its orbit and fired on any insurgents it observed which signalled the vehicles to speed forward and engage with overwhelming firepower. To avoid friendly fire from the helicopters, Koevoet trackers on the ground reversed their hats to reveal bright orange reflective panels. When helicopters were not available, the SAAF dispatched small propeller driven reconnaissance aircraft to try to spot the insurgents or panic them into diving under trees. Koevoet developed several other innovations to enhance cooperation with helicopters. Each team's supply vehicle carried fuel to keep a helicopter overhead longer. Given the lack of prominent landmarks and flat terrain that often disoriented pilots when responding to directions from the ground, a clock code procedure was developed in which the shadows of trees, seen the same way from the air and ground, always represented 12 o'clock.[79] According to veteran South African helicopter pilot Neall Ellis who supported many Koevoet operations and subsequently flew in numerous other conflicts, 'this was helicopter warfare at its most effective'.[80] Of course, the system did not always work perfectly such as during an incident in 1984 remembered by Koevoet tracker Sisingi Kamongo:

> We call the cars and the gunships on standby are summoned. The gunships have a 25-minute reaction time. We start tracking; I am a young tracker able to run on the tracks. The gunship arrives but the pilot is new. He does not know much and he is too far ahead. At one stage, he says over the radio that he has seen movement and we can hear him firing. We follow up and find a dead kudu![81]

Beginning in 1983, given instances where insurgents had escaped pursuit when helicopters were refuelling, the SAAF introduced a Dakota transport aircraft fitted with a 20mm cannon in its cargo door that would fly a wide orbit at around 1000 feet above a Koevoet team. Codenamed 'Dragon', after a similar American aircraft made famous during the Vietnam War, it did the same job as a helicopter gunship but could remain airborne much longer and used loudspeakers as a psychological weapon to broadcast surrender messages to fleeing insurgents.[82] PLAN insurgents considered the effective use of air support to represent the main advantage of Koevoet and army RM teams.[83]

At times the tension of tracking formed the context of brief humorous incidents. In the early 1980s Granger Korff's paratrooper platoon discovered a trail formed by the distinctive chevron pattern footprints of over 100 PLAN insurgents. Although Korff began to nervously wonder if his small unit was capable of engaging this many enemy, 'The one black tracker was stealing the show, so happy to find the spoor that he was smiling and lying on the spoor, making as if he was fucking it. Our troops laughed. He was beside himself and got up and danced a jig on the spoor. We would joke about this display of glee for a long time to come'.[84] During Operation Sceptic in 1980 in southern Angola, South African paratroopers destroyed SWAPO bases and followed the tracks of fleeing enemies. 'On one such track they came across a boot in the sand with patches of wetness leading away from the boot on the ground. When they saw this they all laughed at the individual who had "wet himself" as he fled with one boot"'.[85]

Anti-Tracking

According to a South African journalist who covered the Border War, 'The average PLAN insurgent was often good at field craft, camouflage and anti-tracking techniques, and capable of extreme feats of physical endurance'.[86] The most successful period of PLAN's guerrilla war in northern South West Africa was between around 1976 and 1979 when it heavily infiltrated western Ovamboland and Kavango, and established semi-liberated zones in parts of the dense bush of eastern Ovamboland. The insurgents benefited from recently acquired staging areas in southern Angola and the security forces' use of white conscripts, army reservists and police from South Africa who did not know the environment. In the semi-liberated areas, it appears that it was more common for PLAN insurgents to track and engage South African units that would occasionally enter than for the guerrillas to use anti-tracking to try to avoid contact. At times PLAN conducted impromptu rudimentary training of recruits in the semi-liberated zones but brought them back to southern Angola for more formal basic guerrilla training which included aspects of tracking and anti-tracking. Insurgents assigned to specialist reconnaissance detachments, which scouted infiltration routes and avenues of approach for attacks and guided regular guerrilla groups, received more advanced training in subjects such as concealment, tracking and anti-tracking. PLAN's fortunes changed at the end of the 1970s when the South Africans began to systematically and widely employ primarily African tracking units such as Koevoet and eventually army RM teams routinely supported by airpower. Increasing in frequency and scale, conventional SADF incursions into southern Angola and sponsorship for UNITA further undermined PLAN operations in northern South West Africa and by the early 1980s there were no more semi-liberated zones and the insurgency had been effectively confined to Ovamboland. By around 1980 PLAN, embroiled in the war in Angola, had evolved into a large military force including several conventional mechanized brigades that protected infrastructure in southern Angola and cooperated with Angolan and Cuban forces in the fight against UNITA and its South African supporters. In fact, by the late 1980s several mechanized battalions in Angola integrated PLAN and Cuban troops. PLAN insurgents who continued to infiltrate northern South West Africa received guerrilla warfare instruction at the Tobias Hainyeko Training Centre in Lubango, Angola which included basic tracking and anti-tracking. Specialized PLAN formations such as Typhoon units, formed in 1979 to concentrate on long-range infiltrations such as attacks on the white farming areas to the south, and reconnaissance detachments received more advanced training in concealment and anti-tracking which included extensive practical application and culminated in a 300 kilometre tracking/anti-tracking exercise. Former PLAN members who defected to Koevoet told their new masters that insurgent trainees in Angola had to practice anti-tracking methods even when going to the toilet at night. The students were trained by experienced PLAN insurgents who had survived numerous infiltrations across the border and applied the lessons they had learned in an effort to reduce the mounting casualties inflicted by tracking units like Koevoet.[87]

During the 1980s PLAN cultivated a very high level of anti-tracking expertise. In 1983 a Koevoet commander testified in court that PLAN fighters 'were specifically trained in anti-tracking. They wiped out tracks, retraced tracks, changed footwear ... They have the ability to vanish without trace'.[88] PLAN reconnaissance specialist Ekandjo remembered how he escaped from captivity by South African security forces in 1986 and to avoid recapture 'I had to employ some manoeuvres and guerrilla warfare tactics by walking on dry grass and hard ground to ensure that the enemy would not trace my footprints'.[89] SWAPO insurgents used a multitude of anti-tracking measures some of which were highly imaginative. They wore boots with layers of removable soles that would leave different footprints, wore plastic bags on their feet to make spoor less visible, alternated between walking with boots and bare feet, shuffled along wire fences to avoid touching the ground, walked on fallen leaves or rocks, stepped in each others' footprints, lifted grass bent by their movement, used a stick with a cloth tied to the end or a tree branch to erase tracks, sprinkled water on their footprints to mimic rain which made them appear older, enlisted civilians to destroy their tracks, ignited bush fires to destroy spoor and distract trackers, suddenly changed direction, walked in circles to come back over the spoor of their pursuers, walked on tarred roads when available and walked toward the sun to blind trackers which was also a good way to set an ambush. As in Rhodesia, PLAN units bomb-shelled to make it almost impossible for each individual to be tracked. Larger units of around 100 often dispatched small groups in different directions and at different times, and these small groups further split up into individuals who would rendezvous at a prearranged location. When pursued, SWAPO insurgents tried to increase their speed and endurance by injecting themselves with drugs and commandeering horses, bicycles and motor vehicles from local communities. Knowing the direction their pursuers were coming from enabled insurgents to plant landmines on their tracks, first done in 1971, and set ambushes though security force superiority in firepower, armour and mobility usually made this a desperate and even suicidal measure.[90] In 1987 Koevoet tracker Kamongo, who would eventually lose the use of his legs because of a landmine injury, was among several teams following insurgent spoor: 'We've moved less than 100 meters when there is an ear-splitting BOOM! ... The experienced SWAPO have played their trump card. The POMZ anti-personnel mines they planted have caused devastation. In the sand and among the bushes lie 13 wounded'.[91] One veteran and daring SWAPO insurgent survived numerous Koevoet pursuits by staying just 100 meters ahead of the trackers at which distance they could not spot him in the bush and aircraft flew too far forward to detect him.[92] Sending their younger colleagues forward to make spoor and likely die, some practised PLAN fighters lay still in thick bush and let security force trackers and vehicles pass by them. There were also cases of guerrillas moving just 50 meters ahead of a Koevoet team and pretending to point at spoor with sticks, a Koevoet habit, to deceive circling helicopter spotters into thinking they were security force members. Indeed, some of the most experienced and skilled Koevoet trackers specialized in deciphering PLAN anti-tracking techniques and were only deployed to solve difficult problems. Among these was a former PLAN anti-tracking instructor who had been captured and changed sides but lack of trust in him meant he was kept in a vehicle and brought out only when necessary.[93] Of course, anti-tracking could also be used more aggressively. In early April 1989 PLAN Commander 'Communist' Ambambi, a recent graduate of military college in Yugoslavia who 'liked to confuse his pursuers with the footprints of his men', used anti-tracking to lead an SADF mechanized force into a company ambush:

> Six kilometers from the guerrilla ambush the enemy soldiers came across their footprints. Knowing that the enemy would follow their tracks, the guerrillas had by design made numerous and confusing tracks: the same men doubled back on their own footprints and crossed and re-crossed them again and again in different directions and finally laid an ambush along this trail. That manoeuvre caused the already confused enemy perplexity bordering on panic. They could not work out which of the prints were the most recent and, above all, which ones to follow. They would choose one formation of footprints, then come across another set leading in a different direction, abandon the former and follow the new ones, only to detect more footprints facing in the opposite direction and again turn back to follow them.[94]

Insurgents were most active from November to April as summer rains obliterated their footprints, large and shallow depressions in the ground called 'shanas' filled with water which could not be tracked through, growing vegetation provided cover, mud hindered cross country driving by security force vehicles, and more drinking water was available in the bush. Security force members referred to this busy period as the 'summer games'. In 1981 SWAPO announced that 'Despite usually having to restrict military activity to the rainy season in order to render enemy trackers ineffective, the level of sustained campaigning has always been higher than South African propaganda has cared to admit'.[95]

As with the Rhodesian conflict, South African cross-border operations often reversed the roles of tracker and tracked. A white veteran of the SADF described how his patrol was ambushed in Angola and pursued by 'natural born trackers'.[96] In 1977 32 Battalion, made up mostly of black Angolans, formed a Recce Wing which specialized in using anti-tracking to enter Angola covertly to locate PLAN bases and guide larger attack forces. Infiltrating Angola to gather intelligence and harass PLAN, 32 Battalion platoons tried to avoid detection by anti-tracking such as going barefoot, wearing insurgent footwear and walking ahead of cattle to eliminate their tracks, and particular care was taken while establishing hidden supply cashes which could not fall into enemy hands. Since PLAN personnel were less concerned about anti-tracking inside Angola and prints from their distinctive chevron pattern boot soles were easily observed on trails, 32 Battalion patrols used tracking and back-tracking to locate their bases or set ambushes on regularly used routes or water sources. If 32 Battalion troops observed the tracks of a man and a dog crossing from Angola to South West Africa, they suspected that PLAN had sent a civilian sympathizer to scout an infiltration route as it was a favourite tactic to take a dog to sniff out security force positions. In these cases 32 Battalion personnel back-tracked the civilian and dog, distinguishable as a pair from other civilian tracks, to locate the community they came from inside Angola and establish an observation post to wait for insurgents.[97] Of course, anti-tracking did not always work, particularly if the unit consisted of more than a few people and was moving quickly. SADF paratrooper Korff recalls that in the early 1980s after a running gun fight with the Angolan army inside Angola, his fleeing platoon tried unsuccessfully to conceal their trail. 'The black tracker ran last, trying to do a quick job of clearing out our spoor with a branch of leaves but he wasn't having much success'.[98] From the early days of South African Special Forces in the late 1960s, tracking represented an important part of training. By the late 1970s new operators took a three week bush survival course, sometimes taught in a Natal game reserve, and sometimes in the Caprivi Strip, which included tracking and anti-tracking that 'would later in our careers become of life saving importance'.[99] The course was taught by Bushmen trackers supervised by the legendary Sergeant Major Dewald de Beer who had reputedly lived with a Bushmen community for six months and learned their language, had been the only student to achieve an 'A' grade on the Rhodesian Selous Scouts tracking course, and had once tracked a SWAPO group from the front of a speeding

Landrover. In April 1974, as the Portuguese were leaving their colonies, then Staff Sergeant de Beer had been attached to a four man Rhodesian SAS team that attacked an insurgent camp in Mozambique and then pursued the fleeing enemies for a week. 'Throughout their time in Mozambique the team relied heavily on Dewald for his outstanding tracking bushcraft capabilities'.[100] Rhodesian experience appeared important as Ray Godbeer, another Warrant Officer who ran the bush survival course, had served three years in the RLI and six in the Selous Scouts.[101] Recce operators studied 'spoor reports' collected by their small patrols to determine the size, movement patterns and potential intent of PLAN forces in areas of northern South West Africa and southern Angola. In one instance around Nkongo, South West Africa during the early 1980s, Recce operators had determined that PLAN forces moved only in two man teams but 32 Battalion continued to employ their minimum six man patrol which resulted in their being detected, tracked and ambushed by PLAN along with a 15 man Recce pseudo team. As PLAN became increasingly skilled at camouflaging their bases in Angola, South African Recces copied the Rhodesian two man reconnaissance/sabotage team that could operate undetected inside enemy territory for long periods. With further anti-tracking training, these small teams obsessed with minimizing the impact of every step which meant they moved only five to seven kilometres per day and did not move at night when it was almost impossible to anti-track. They were frequently sent to confirm if a base detected by aerial photography was actually inhabited and, if so, summon an attack force. According to highly decorated operator Greeff, 'Should SWAPO detect any foreign spoor, they would evacuate the base and the attack would be a lemon'.[102]

Security Force Tracking in South Africa

As its protective ring of adjacent regional allies collapsed in the 1970s and 1980s, apartheid South Africa itself became the scene of military oriented tracking and anti-tracking. In the 1960s and early 1970s South Africa was associated with the white minority regime in Rhodesia and the Portuguese colonial rulers of Angola and Mozambique, it administered South West Africa and economically dominated the small, impoverished and newly independent African states of Botswana, Lesotho and Swaziland. This began to change with the departure of the Portuguese from Africa in 1974 and the independence of Zimbabwe in 1980. There was also growing sympathy for African liberation movements within Botswana's government which began to exercise more autonomy by using revenues from its emerging diamond industry to establish a defence force in the late 1970s. When the anti-apartheid ANC and Pan-Africanist Congress (PAC) went to exile in the 1960s, they were based in recently independent African countries like Zambia and Tanzania that were geographically removed from South

Africa and therefore, it was very difficult for their insurgents to infiltrate home territory to pursue the armed struggle. At that point, and with Eastern Bloc support, military training for the ANC's armed wing Umkhonto we Sizwe (Spear of the Nation or MK) consisted mostly of political ideology, technical weapons instruction and parade ground drill with little emphasis on field operations or bush craft skills such as tracking or anti-tracking. This is understandable as most recruits and commanders were from urban or semi-urban backgrounds, and the movement became involved in conventional warfare in Angola and both urban and rural guerrilla operations. However, it appears MK operatives in South Africa in the 1960s, like PLAN fighters some years later, tried to use pepper and curry powder to throw off security force tracking dogs. Although the PAC tried and failed to infiltrate South Africa through Mozambique in the late 1960s and Botswana in the early 1970s, the organization became engulfed in internal conflict which prevented it from launching effective operations until the end of the 1980s. After Portuguese decolonization in the middle 1970s, the ANC launched military training in Angola where it was allied with the ruling MPLA and committed troops to the war against UNITA to give its fighters combat experience, counter disciplinary problems in its camps and contribute to the regional struggle against apartheid South Africa. In Angola MK organized new instructional programs including basic training, communications, intelligence, engineering and artillery, and some members were sent to Cuba for special training in rural guerrilla warfare though it is unclear exactly what this involved. After the 1976 Soweto Uprising which reinvigorated the internal anti-apartheid struggle, many young black South Africans left their country and swelled the ranks of the exiled ANC and MK. In the late 1970s and early 1980s hundreds of MK operatives infiltrated South Africa to build covert networks and engage in limited military missions called 'armed propaganda' which targeted police stations, railway lines and other state institutions to motivate mass action campaigns. During the early to mid-1980s the ANC shifted from these limited propaganda operations to pursuing a protracted people's war in South Africa which focused on targeting vulnerable black municipal officials and planting landmines in white farming areas. This meant increased weapons smuggling and infiltration of guerrillas from Botswana, Zimbabwe and Mozambique. However, these avenues of approach were narrowed in 1984 when Mozambique's FRELIMO government, under pressure from South African sponsored rebels, agreed to expel ANC activists.[103] During the 1980s the SAP and security forces from the recently created black homelands such as Venda and Bophuthatswana patrolled South Africa's borders and called in police or army reaction units to pursue infiltrators. When white farmers near the borders found suspicious footprints or there was a landmine incident, the police would dispatch 14 man reaction teams consisting mostly of white members some of whom had advanced tracking training, mounted in four-wheel drive trucks to investigate and track insurgents. In this situation, the guerrillas were usually a day or two ahead of security force tracking teams which tried, like in South West Africa, to use vehicle mounted advance groups to catch up but this was difficult in the bush-covered and hilly environment. It proved more effective to trap insurgents by using teams dispatched to different positions by motor-vehicle and helicopter. Some ANC fighters were skilled in anti-tracking and had obviously been trained to some extent, and their lead time after crossing the border meant they had the opportunity to employ these measures. The captured diary of an MK fighter who was part of a small group that had secretly crossed from Swaziland into the Natal bush in 1984 displays a keen awareness of the importance of concealing their own tracks and observing others' tracks to learn about security force movements. Sometimes the police used tracking dogs but, as in South West Africa, they were of limited value as they became exhausted in the heat of the day. Although the military history of the apartheid wars focuses on events in Angola and Namibia, many MK guerrillas who crossed from neighbouring countries and some security force members were killed in these tracking operations in northern South Africa.[104]

Game-Keepers and Anti-poaching

The influence of state game-keepers in South African counter-insurgency in South West Africa was relatively limited. Some National Parks trackers were employed by security forces but they were few because the large Etosha National Park, with its massive and open pan, signified more of an obstacle to insurgents than a sanctuary. National Parks in South Africa and South West Africa were used as tracker training areas, and some park staff became instructors. After the independence of Namibia in 1990 and the advent of democracy in South Africa in 1994, some veterans of both sides of the conflict put their military tracking expertise to work in the conservation sector. By 2001, the Namibian Ministry of Environment and Tourism had employed 2000 ex-combatants with many in anti-poaching units and other PLAN veterans joined the para-military Namibian Police Special Field Force.[105] In the 1990s accomplished Special Forces operator Greeff left the South African military to become a game ranger in South Africa's Kruger National Park where he trained personnel in combat tracking and ambush techniques. It is rumoured that Greeff's aggressive methods were highly effective in killing suspected poachers but also led to his departure from the park after just a few years as the mounting body count seemed embarrassing to post-apartheid authorities. Along with other Border War veterans, he became a private consultant and instructor for increasingly militarized anti-poaching units in National Parks and private game reserves across the region and as far away as the Democratic Republic of Congo (DRC).[106] With an increase in rhino poaching by gangs operating across the border in Mozambique, Kruger National Park deployed units of the post-apartheid South African National Defence

Force (SANDF) to patrol its boundaries and in 2012 a former SADF general was hired to direct anti-poaching operations. However, SANDF had largely abandoned counter-insurgency training as it brought up unpleasant memories of the apartheid era which meant it lost much of the tracking expertise built up over the late 1970s and 1980s.[107] Reminiscent of the days of the apartheid wars, poachers took on the role of insurgents trying to infiltrate South Africa from neighbouring countries. In 2013 Greeff, back training rangers in Kruger National Park, told a reporter that 'I was able to take my passion for the environment and mix it with my military experience. We train for guerrilla warfare here. A poacher is a criminal and a saboteur that acts as an insurgent. We therefore train the field rangers to conduct anti-insurgency operations'.[108] In this context, Bushmen from remote areas continue to be sought after as trackers. In 2013 Namibian police arrested a former SADF colonel for alleged human trafficking after he recruited a dozen Bushmen from the Bwabwata National Park in the Caprivi Strip to work as trackers for his security company Wildlife Investigation and Protective Services (WIPS) which specializes in countering rhino poaching and providing farm security in South Africa.[109]

South Africa's counter-insurgency in South West Africa also influenced the development of military-style anti-poaching operations in neighbouring Botswana. By the late 1980s, lightly armed and unprepared Botswana police and wildlife officers were struggling to counter mounting cross-border poaching in the country's northern national parks and concession hunting areas. In 1987 Sandhurst graduate Major Otisitswe B. Tiroyamodimo, commander of the Botswana Defence Force (BDF) Commando Squadron, visited South West Africa disguised as a Botswana wildlife official pretending to engage in talks on conservation issues but in reality he was there to recruit Botswana citizens who were working as trackers in the South African security forces. Subsequently, around 50 such trackers were incorporated into the BDF Commando Squadron which set up a headquarters in the northern town of Maun and quickly embarked on a program of sending four to six man anti-poaching patrols deep in the vast wilderness of northern Botswana. While most of the trackers recruited in South West Africa initially lacked the educational requirements to formally enrol in the BDF and were hired as civilian contractors, some eventually acquired these qualifications and were enlisted as regular soldiers. Three days after the patrols were launched the Commando Squadron experienced the first of many fire-fights with well-armed poachers and within a year the number of poached elephants in the area had dropped by more than half. In 1989 the BDF broadened its anti-poaching campaign by deploying conventional infantry companies in the north, and eventually other parts of the country, supported by vehicles and helicopters which represented a transition from aggressively hunting poachers to deterring them with a show of force. In addition, the BDF incorporated wildlife familiarization into its regular training program to educate its members about the value of conservation and teach them how to deal with animal encounters during anti-poaching patrols which continue to this day. Soldiers from conventional units also took military tracking courses. Indeed, the BDF's Commando Squadron has maintained its combat tracking expertise and recently trained United States Special Forces members in this skill.[110]

Conclusion

Just as the flat, semi-open terrain and sandy soil of parts of the Namibia/Angola border facilitated the persistence hunting of traditional hunter-gatherers, it also enabled late twentieth century South African security forces to develop a highly successful approach to counter-insurgency tracking in which wheeled armoured vehicles allowed indigenous trackers to move fast and leapfrog, and protected them from ambush, and air support arrived at a critical moment. The most successful of these operations were not conducted by conventional units with attached special-ist tracker teams but by large combined arms reaction units like Koevoet and 101 Battalion that incorporated tracking into almost everything they did. While early efforts by the police and army focused on employing trackers from the Bushman minority, the security forces' real success in tracking happened once they mobilized the majority Ovambo whose historic way of life was much less romanticized but also produced excellent trackers. This was facilitated by the pragmatic and para-military approach of the police which allowed for individual experimentation, a change in military culture within the SADF which began to utilize black military manpower, and an expensive military expansion that invested in mine-resistant vehicles and various aircraft. The use of fast horse and motorcycle mounted tracking units and scent tracking dogs was confined to certain areas that were somewhat open or where insurgent infiltration was less intense. With daily pursuit and evasion, the conflict in northern South West Africa and southern Angola in the 1980s turned into a tracking war with both sides continually refining their methods. According to Koevoet veteran Arn Durand, his unit focused on tracking while SWAPO learned to anti-track which led to Koevoet developing techniques to 'anti-anti-track' with the insurgents then learning to 'anti-anti-anti-track'.[111] However, it must be remem-bered that the South African emphasis on tracking represented a central feature of an ultimately unsuccessful counter-insurgency strategy that resulted in withdrawal and an internationally negotiated settlement which allowed SWAPO to gain power in an independent Namibia. By prioritizing the tracking and killing of insurgents, which gave the illusion of military victory based on a high number of enemy deaths, much less emphasis was placed on winning civilian 'hearts and minds' and making real political reforms to render insurgency irrelevant. Lack of cooperation between military and police also became a problem. Nevertheless, the South African com-bat tracking innovations during the Border War could have been framed within the context of a more comprehensive and effective counter-insurgency program which means they may offer important lessons for future practitioners.

CONCLUSION

During Africa's late twentieth century decolonization wars, tracking became an important skill mobilized by both state security forces and insurgents. To some extent this was informed by established stereotypes that associated tracking skill with specific marginalized minorities such as the Ndorobo in Kenya, Shangaan in southeastern Zimbabwe and Bushmen in the Kalahari region, and the use of tracking in nineteenth century colonial warfare which created colonial military tracking heroes such as R. S. S. Baden-Powell and Frederick Russell Burnham. There was a clear evolution in the use of tracking in the counter-insurgency campaigns carried out in white minority dominated Kenya in the 1950s, Rhodesia (Zimbabwe) in the late 1960s and 1970s and South West Africa (Namibia) in the 1970s and 1980s. While security force officials were influenced by previous or ongoing campaigns in other places, the application of tracking to counter-insurgency was largely determined by local geography, technology and colonial culture. Of course, it must be remembered that tracking represented an important part of just one element of counter-insurgency; that of engaging the guerrillas themselves. It had little to do with attempts to win the 'hearts and minds' of the civilian population though it could be said that effectively locating guerrillas would prevent them from subverting broader society.

The usual view of British counter-insurgency in Kenya during the 1950s is that the failure of large and clumsy security force sweeps eventually gave way to more effective small unit operations involving pseudo-teams and tracker combat teams (TCTs). In fact, tracking was practised right from the beginning of the emergency and preparations for the formation of specialized tracker units began very early in the conflict. State game-keepers such as Second World War veteran Rodney Elliott were particularly influential in recruiting and promoting the use of indigenous trackers. Kenya's local hunting culture characterized by a white hunter who did the shooting and his faithful black tracker who found the game was important in shaping security force tracking during the Mau Mau Rebellion. A Tracking School, staffed by white game-keepers or hunters and their black assistants, was created that tested putative African trackers and gave them relevant military training and showed European leaders how best to utilize them.

Most of the indigenous trackers initially recruited by the security forces were from groups other than the Kikuyu, the main community involved in Mau Mau, particularly the marginalized Ndorobo minority who had a long association with white game-keepers and hunters and supposedly warrior peoples like the Kamba. However, as the emergency developed, more Kikuyu including loyalists and turned insurgents were put to use as security force trackers as they knew the forest environment and Mau Mau anti-tracking practices. In Kenya security force trackers were mostly Africans with local expertise who were supervised by white superiors with hunting experience and/or special training. Security force tracking in Kenya was based on the age-old method of persistence hunting with police or military patrols searching for insurgent tracks and then trying to catch up with them on foot which was very difficult. Pioneered by white farmer-hunter Venn Fey accompanied by his lifelong Ndorobo tracker Gichimu, the main combat tracking innovation in Kenya involved the grouping of several tracker combat teams to work in a specific area under a single commander. In this way, persistence hunting was improved upon by having a fresh team take over a pursuit from an exhausted one and teams were dispatched, sometimes by motor vehicle along specially cut forest corridors, to areas the insurgent prey were likely to pass through. Trained dogs were important to security force tracking in Kenya as scent tracking dogs provided by the Royal Veterinary Corps were acclimatized at the Tracking School and each TCT was supposed to have two dogs; one for scent tracking and one for early warning. It is important to realize that the new technology of helicopters was not used by British security forces in Kenya during the Mau Mau war and it would have been difficult to employ them in the forests where tracker teams operated. It is also important to remember that counter-insurgency operations in Kenya focused on sizeable but very specific forest areas around the Aberdares and Mount Kenya, and that the insurgents lacked a foreign sanctuary or external sponsor. While British counter-insurgency practitioners in Kenya and Malaya influenced each other during the 1950s, it appears that the approach to combat tracking developed in East Africa was applied by the Jungle Warfare School in Asia and influenced the development of combat tracking by other forces such as the Americans in Vietnam.

The use of tracking by Rhodesian counter-insurgency forces was shaped by the white community's frontiersman self-imagine and fear of mobilizing armed black soldiers. Influenced to some extent by events in Kenya and Malaya, Rhodesian security force interest in tracking began on the eve of the insurgency with game rancher Alan Savory's personal crusade that resulted in the creation of the Tracker Combat Unit (TCU) and SAS tracker teams, and the separate and simultaneous formation of the Police Anti-Terrorist Unit (PATU). Throughout the war state game-keepers played a key role in counter-insurgency forces by providing African trackers under European leadership and specialist training. Security force tracking was highly successful during the first phase of the conflict in the late 1960s and very early 1970s when Zambia based insurgents, mostly from urban backgrounds, lacked expertise in anti-tracking measures and operations took place in the remote and sparsely populated Zambezi Valley where it was generally easy to observe spoor. Dissatisfied with National Parks and civilian trackers who were mostly black, the Rhodesian Army opened its Tracking Wing in 1970 that trained mostly white military personnel. Unlike in Kenya, the whites would do the tracking itself and not just manage black trackers. The Rhodesian Army quickly expanded its tracking expertise and training, and assisted allied Portuguese forces in Angola and Mozambique who lacked personnel with these skills. Perhaps the most important Rhodesian innovation in combat tracking was to use helicopters, acquired by the Rhodesian Air Force during the federal period, to leapfrog tracker teams along suspected insurgent trails in anticipation of catching up with and engaging the enemy. This was facilitated by semi-open terrain in the Zambezi Valley which enabled aerial observation and helicopter landings. Furthermore, given some failed experiments, the Rhodesian Army avoided using tracker dogs as it was believed scent evaporated in the hot climate. The police seemed more pragmatic about tracking and always employed part-time African trackers mostly from marginalized minorities, often used tracker dogs and did not establish a formal tracker training centre. Tracking became much less useful for the Rhodesian security forces during the second phase of the war in the 1970s when ZANLA insurgents established staging areas in neighboring Mozambique to infiltrate the vast, populated and sometimes forested east of Rhodesia, and the guerrillas who were increasingly from rural backgrounds began practicing effective anti-tracking techniques. Despite the popular image of the Selous Scouts as a tracking unit, the Rhodesian Army emphasis on tracking declined in the mid to late 1970s as the overwhelmed security forces responded with pseudo-gangs, air mobile reaction forces and protected villages. At the same time, the army tried to mobilize the famous tracking skills of the Shangaan minority but, similar to the broader expansion of black combat units, it was too little and too late.

In South West Africa during the late 1970s and 1980s environment, technology and security force culture combined to greatly expedite counter-insurgency tracking. In contrast to events in Kenya and Rhodesia, the police took the lead in applying tracking to warfare. When the insurgency began in the late 1960s the South African Police (SAP) built on their long relationship with local Bushmen minority trackers to pursue guerrillas from the majority Ovambo community. Informed by its experience in Rhodesia's Zambezi Valley and unfolding events in South West Africa, the SAP established a counter-insurgency training centre and began tracker training in South Africa during the early 1970s. The police realized fairly early that relying on minority trackers would not sufficiently challenge the mounting insurgency and began forming home guard units among the majority Ovambo people which was also the main recruiting pool for SWAPO. Beginning in the late 1970s, a police unit called Koevoet commanded by career police officer Hans Dreyer and developed by innovative officers like Eugene de Kock used highly flexible tracker platoons mounted in mine-resistant armoured trucks to find and follow insurgent spoor, and eliminate the insurgents. The teams were led by experienced white policemen but consisted mostly of skilled Ovambo trackers and former insurgents. Persistence hunting was facilitated by having the vehicles drive next to the trackers who were spelled off and given water, leapfrogging was done by vehicle mounted teams that drove forward along an insurgent trail and effective procedures were developed for cooperation with South African Air Force helicopter gunships called in during the last phase of a pursuit. All this was made possible by the mostly open or semi-open terrain of northern South West Africa. In the initial period of insurgency in the early 1970s, the SADF dispatched entirely white South African units mostly consisting of young conscripts to the operational area. The departure of the Portuguese from neighbouring Angola and the failure of the South African invasion of that country in the middle 1970s brought some black Angolans into the hitherto all white SADF which was beginning to recognize the need to employ African soldiers in counter-insurgency. The first dedicated SADF trackers in South West Africa were Bushmen soldiers the nucleus of whom originated from Angola. These were seconded as trackers to conventional units and in the late 1970s the Bushman battalion, not initially formed as a tracking unit, began using mechanized reaction teams to track insurgents. By the middle 1980s the SADF had formed its own version of Koevoet called 101 Battalion and similarly based on Ovambo personnel. The SADF also formed highly mobile horse and motorcycle mounted units to patrol a cleared strip along the border and track insurgents but these were vulnerable to ambush. While the South African security forces employed tracker dogs more systematically than the Rhodesians, there were overgrown areas where they were not used because of the likelihood of ambush. Unlike Kenya and Rhodesia, state game-keepers did not play a major role in active counter-insurgency though some national parks in South Africa were used as security force training areas and Peter Stark, former head warden of Etosha National Park, became the chief riding instructor at the SADF equestrian school. SWAPO responded to South African tracking by founding its own anti-tracking school in Angola, utilizing many highly imaginative anti-tracking methods, planting landmines and ambushes in the path of pursuers, retreating across the Angolan border when necessary and flooding the border area with guerrilla infiltrators.

Counter-insurgency forces in late twentieth century East and Southern Africa tried to utilize the superior senses and physical abilities of domestic animals. An important feature of security force tracking in 1950s Kenya was that human visual trackers were augmented by trained scent tracker dogs and patrol dogs that gave early warning of ambush. While dogs proved not very effective in the campaigns fought in the heat of Rhodesia and South West Africa, it is likely that the colder conditions of the high altitude Kenyan forests meant that scent remained on the ground longer and the British army had established expertise with dogs. It is also worth remembering that humans' ability to track visually is related to the experience of persistence hunting which animals like dogs, unable to regulate body temperature by sweating, cannot do. Although in some circumstances tracker dogs might hinder pursuit by obliterating spoor or collapsing from exhaustion, they offered the advantage of being able to track at night which was impossible for human visual trackers. The great speed of a horse and the high vantage point of a rider meant that mounted infantry was used in all three campaigns as well as by the Portuguese in Angola. Although horses were of limited use in the high Kenyan forests and usually used to patrol the more open northern frontier, security forces in Rhodesia and South Africa established permanent mounted infantry units though their vulnerability to ambush restricted their use. Insurgents also used animals such as civilian owned dogs to check for concealed enemies and stolen horses to facilitate a speedy escape from trackers.

To expedite the tracking of insurgents, counter-insurgency forces in Kenya, Rhodesia and South West Africa altered the environment. In 1950s Kenya British army engineers cut long tracks into the forest which were used by security forces to more easily head off Mau Mau groups that were being pursued by tracker teams. The forest tracks also enabled the British to establish support bases used by tracking teams and other units within the Mau Mau sanctuary. In 1970s Rhodesia the state constructed a narrow fenced and cleared strip along its eastern border with Mozambique and bulldozed large swaths of bush in so called 'Free Fire Zones' which were regularly patrolled for signs of insurgent infiltration. During the 1970s and 1980s South African forces in South West Africa created a much wider band of depopulated and defoliated territory along its northern border with Angola which was similarly patrolled for insurgent spoor that, if discovered, would be quickly followed up by a variety of reaction units. In Rhodesia and South West Africa, these border cordons were regularly graded by machines so that footprints would be more plainly visible.

Given that insurgents interact with and target civilians, both military and police forces are usually involved in counter-insurgency warfare. In these circumstances, police usually have investigative capabilities and legitimacy with the public that the military lacks, and the military usually has much greater combat effectiveness such as firepower and mobility. In all three cases discussed in this book, the army and police approached counter-insurgency tracking in different ways. The army always established formal tracking schools that fed personnel to dedicated combat tracking units. The 1950s Tracking School (absorbed as a distinct wing into the East Africa Battle School) in Kenya supported the British

Army TCTs, the 1970s Rhodesian Army Tracking Wing served units like the TCU and SAS before its absorption into the Selous Scouts, and South West Africa's SWAPSES operated its own tracking school during the 1980s. On the other hand, the police in these conflicts usually approached tracking in a more ad hoc manner and often sought assistance from civilian trackers. The police usually avoided creating dedicated combat tracking schools preferring to incorporate such skills into wider counter-insurgency training programs. The Kenya Police created small tracker teams led by white professional hunters in direct response to insurgent stock theft in specific areas. In Rhodesia, PATU employed trackers but was not a specialist tracking unit in the same way as the army's TCU or tracking teams fielded by units like the SAS or RLI. In South West Africa, Koevoet was a large police counter-insurgency unit that relied on the existing tracking skills of its Ovambo members which were adapted on the job to counter-insurgency operations and combined with many other elements of fast and flexible investigations and pursuits. It is also important to note that aspects of air power represented an important element of counter-insurgency tracking involving both army and police in all three campaigns.

The involvement of state game-keepers in decolonization era counter-insurgency operations, where they were involved in tracker training and actual tracking operations had a major impact on the development of post-colonial military-style anti-poaching in East and Southern Africa. The work of Bill Woodley and other Mau Mau conflict veterans in forming the Voi Field Force became a model for anti-poaching in Kenya and other African countries up to the present. Although independent Zimbabwe enacted very different conservation policies than Kenya, National Parks officials and former soldiers with Rhodesian counter-insurgency experience became central to the mounting of military anti-poaching operations after 1980. State game-keepers were only peripherally involved in South African counter-insurgency in South West Africa but after 1990 many veterans of that conflict joined anti-poaching campaigns across the region.

What can contemporary forces learn from this history of counter-insurgency tracking in Africa? Over the past few years the Ugandan People's Defence Force (UPDF) has been hunting the infamous Lord's Resistance Army (LRA) in the vast and forested tri-border area of the Central African Republic (CAR), Democratic Republic of Congo (DRC) and South Sudan which is roughly the same size as California. Since its emergence in northern Uganda in the 1980s, the LRA has consistently brutalized communities and kidnapped children who became fighters, porters and sexual slaves. In 2012, given belated international outrage over LRA atrocities against civilians and International Criminal Court indictment of its leaders, the United States dispatched 100 Army Special Forces advisors to the region to assist the UPDF in its hunt though they are not meant to be directly involved in operations. At that time the UPDF could operate against the LRA in

CAR and South Sudan but not in DRC or the nearby Darfur region of Sudan. In this remote region borders are often meaningless and the LRA has freely crossed them. With 800 soldiers in CAR and 700 in South Sudan, the UPDF maintained ten support bases manned by 50 to 150 troops each from where short range patrols were mounted. Some 20 tracking teams of 20 to 40 men each operated deeper in the bush for longer periods supplied every ten days by helicopter or truck along the few existing roads. Like previous counter-insurgency forces, the Ugandan tracking teams also employed former LRA members who knew the group's methods and habits. During the rainy season of July to September the main supply route from Uganda through South Sudan into CAR was mostly impassible. Although Ugandan soldiers have been reported as very physically fit, the extent of their tracking training and expertise is unknown. Two rented Mi8 transport helicopters paid for by the United States and flown by civilian pilots were responsible for both logistical and troop transport flights, and were contractually unable to remain outside their main base overnight which limited their operational use. UPDF intelligence on the LRA came mostly from local civilians as aerial reconnaissance from manned and unmanned American aircraft was unable to see through the area's dense forest canopy. Ugandan tracking teams, communicating with their headquarters by radio or satellite phone, lacked the mobility to respond to the latest intelligence. The Ugandan trackers reported going weeks in the forest without seeing a sign of LRA activity, rarely catching up to their prey when on their trail and often tracking a group for several days only to discover they were following poachers or bandits.[1]

Historical examples clearly show that in this situation Ugandan forces had very little chance of successfully hunting the LRA. As the Mau Mau Emergency illustrated, in a forest area security force trackers on foot can rarely catch up with determined insurgents who are more than a few hours ahead. There were too few Ugandan troops to cover a massive area where forests and very poor roads inhibit motor transport. Furthermore, there were not enough helicopters to keep even these small forces supplied in the field or respond to insurgent sighting let alone try to copy the Rhodesian practice of leap-frogging trackers or deploying cut-off groups. The sanctuary provided to the LRA by international borders that the UPDF was not supposed to cross was reminiscent of periods of the Rhodesian and South West African campaigns. The mobility problem was recognized and in March 2014 the United States deployed four CV-22 Ospreys, fixed wing aircraft that can transform into helicopters and carry 24 men each, along with refuelling aircraft and 150 support personnel to the region to help the Ugandans hunt for the LRA. However, these aircraft were inexplicably withdrawn after less than a month despite the capture of a junior LRA leader by Ugandan forces in CAR during that time.[2] To eliminate the LRA it will almost certainly be necessary to deploy many more troops with effective logistical support, transportation and communications. Limited security force manpower and particularly a relative shortage of tracking specialists posed a major problem for Rhodesian counter-insurgency in the 1970s. Although the lessons of the British Army's forest operating companies and TCTs in 1950s Kenya may be worth applying to this situation as the environment is somewhat similar, the Mau Mau forest haunts were much smaller than the current LRA zone and did not include international borders. While it is not known to what extent Ugandan forces employ local trackers, history shows that indigenous trackers with relevant training, support and supervision work best in hunting elusive guerrillas. As the wars in 1950s Kenya and 1980s South West Africa showed, these indigenous trackers need not just originate from minority ethnic groups stereotyped as gifted trackers and hunters. In general, the most effective trackers in these conflicts were from the majority groups from which the insurgents also originated. The practical experience of successful South African reaction units like Koevoet and 101 Battalion in South West Africa may not be that useful in the LRA case as extensive use of cross-country motor vehicle transport is very limited in the forested CAR/DRC/South Sudan border area. Perhaps the most important historical lesson here is to look at the Rhodesian and South African use of helicopters in tracking operations. A small fleet of perhaps several dozen medium helicopters each capable of carrying around ten soldiers should be positioned at a series of forward operating bases throughout the LRA zone to provide short-notice reaction to intelligence. Improving local ground transport and communications infrastructure such as roads will also be important though a dramatic improvement is probably not realistic. Obviously the expense of such a campaign would be beyond the means of most African states and, if implemented, would have to work in conjunction with programs meant to encourage LRA surrenders such as ongoing radio broadcasts, amnesty offers and meaningful rewards for useful information. While the hunt for the LRA has been disrupted by recent civil wars in CAR and South Sudan, it is hoped that it will eventually benefit from the lessons of combat tracking history.

WORKS CITED

Adams, M. and C. Cocks, *Africa's Commandos: The Rhodesian Light Infantry* (Solihull, UK: Helion, 2013).

Alexander, E. G. M., 'The Cassinga Raid' (MA Thesis, University of South Africa, 2003).

Alexander, J. E., *Narrative of a Voyage of Observation Among the Colonies of Western Africa, in the Flag-ship Thalia, and a Campaign in Kaffir-Land* (London: Henry Colburn, 1837).

Alexander, J., 'Dissident Perspectives on Zimbabwe' Post Independence War', *Africa: Journal of the International Africa Institute*, 68:2 (1998), pp. 151–82.

Alexander, M., 'South African Airborne Operations', *Scientia Militaria: South African Journal of Military Studies*, 31:1 (2001), pp. 49–82.

Anderson, D., *Histories of the Hanged: The Dirty War in Kenya and the End of Empire* (London: Weidenfeld and Nicolson, 2005).

Anderson, G. H., *African Safaris* (Edmonton: Safari Press, 2000).

Anderson, R., *The Forgotten Front: The East African Campaign, 1914–1918* (Stroud, UK: Tempus, 2004).

Baden-Powell, R. S. S., *Aids to Scouting for N.C.O.'s and Men* (1899; Aldershot: Gale and Polden, 1915).

—, *Cavalry Instruction* (London: Harrison and Sons, 1885).

—, *Scouting for Boys* (London: Pearson, 1908).

—, *The Matabele Campaign 1896* (London: Methuen, 1900).

Bagshawe, F. J., 'The Peoples of the Happy Valley (East Africa) The Aboriginal Races of Kondoa Irangi Part II: The Kangeju', *Journal of the Royal African Society*, 24:94 (January 1925), pp. 117–30.

Baines, G., *South Africa's Border War: Contested Narratives and Conflicting Memories* (London: Bloomsbury, 2014).

Baker, W. D., *Dare to Win: The Story of the New Zealand Special Air Service* (Melbourne: Lothian, 1987).

Baldwin, W. C., *African Hunting From Natal to the Zambesi* (London: Richard Bentley, 1863).

Bartlett, F., *Shoot Straight and Stay Alive: A Lifetime of Hunting Experiences* (Johannesburg: Rowland Ward Publications, 1994).

Basson, N. and B. Motinga, *Call Them Spies* (Windhoek, Namibia: Africa Communications Project, 1989).

Battistoni, A. K. and J. J. Taylor, 'Indigenous Identities and Military Frontiers: Reflections on San and the Military in Namibia and Angola, 1960–2000', *Lusotopie*, 16:1 (2009), pp. 113–31.

Bax, T., *Three Sips of Gin: Dominating the Battlespace with Rhodesia's Elite Selous Scouts* (Solihull, UK: Helion and Company, 2013).

Beattie, H., 'The Maori as a Tracker and Signaller', *The New Zealand Railways Magazine*, 14:5 (August 1939), pp. 40–4.

Beckett, I., *Modern Insurgencies and Counter-Insurgencies; Guerrillas and Their Opponents since 1750* (London: Routledge, 2001).

Beinart, W. and P. Coates, *Environment and History: The Taming of Nature in the USA and South Africa* (London: Routledge, 1995).

Beinart, W. and L. Hughes, *Environment and Empire* (Oxford: Oxford University Press, 2007).

Bennett, H., *Fighting the Mau Mau: The British Army and Counter-Insurgency in the Kenya Emergency* (Cambridge: Cambridge University Press, 2013).

— and R. Cormac, 'Low Intensity Operations in Theory and Practise: General Sir Frank Kitson as Warrior-Scholar', in A. Mumford and B. C. Reis (eds), *The Theory and Practise of Irregular Warfare: Warrior Scholarship in Counter-Insurgency* (New York: Routledge, 2014), pp. 105–24.

Berman, B. and J. Lonsdale, *Unhappy Valley: Conflict in Kenya and Africa* (London: James Currey, 1992).

Binda, A., *The Saints: The Rhodesian Light Infantry*, ed. C. Cocks (Johannesburg: 30 Degrees South, 2007).

—, , *Masodja: The History of the Rhodesian African Rifles and its Forerunner the Rhodesia Native Regiment*, commissioned and compiled by Brigadier D. Heppenstall and the Rhodesian African Rifles Regimental Association, UK (Johannesburg: 30 Degrees South, 2007).

—, *The Rhodesia Regiment: From Boer War to Bush War, 1899–1980* (Alberton, South Africa: Galago, 2012).

Bird, E., *Special Branch War: Slaughter in the Rhodesian Bush Southern Matabeleland, 1976–1980* (Solihull, UK: Helion and Company, 2014).

Blacker, J., 'The Demography of Mau Mau: Fertility and Mortality in Kenya in the 1950s: A Demographer's Viewpoint', *African Affairs*, 106 (2007), pp. 205–27.

Bley, H., *South West Africa Under German Rule, 1894–1914* (Evanston, IL: Northwestern University Press, 1971).

Blundell, M., *So Rough A Wind: The Kenya Memoirs of Sir Michael Blundell* (London: Weidenfeld and Nicholson, 1964).

Bolaane, M., 'The Role Played by Botswana During the Liberation Struggle', Hashim Mbita Project.

Bopela, T. and D. Luthuli, *Umkhonto we Sizwe: Fighting for a Divided People* (Alberton, South Africa: Galago, 2005).

Bothma, L. J. *Buffalo Battalion: South Africa's 32 Battalion: A Tale of Sacrifice* (Bloemfontein: L. J. Bothma, 2007).

Branch, D., *Defeating Mau Mau, Creating Kenya: Counterinsurgency, Civil War and Decolonization* (Cambridge: Cambridge University Press, 2009).

Breytenbach, J., *The Buffalo Soldiers: The Story of South Africa's 32-Battalion, 1975–1993* (Alberton, South Africa: Galago, 1999).

Brown, C. and S. L. Lahren, 'More on Hunting Ability and Increased Brain Size', *Current Anthropology*, 14:3 (June 1973), pp. 309–10.

Brown, M., 'The Tracking School – Nanyuki (1953–54)', *Buffalo Barua: The Newsletter of the Kenya Regiment Association of Europe and North America* (November 2007), pp. 35–40.

Brown, S., 'Diplomacy by Other Means – SWAPO's Liberation War', in C. Leys and J. Saul (eds), *Nambia's Liberation Struggle: The Two-Edged Sword* (London: James Currey, 1995), pp. 19–29.

Brown, T., *The Tracker* (New York: Berkeley Books, 1978).

—, *The Science and Art of Tracking* (New York: Berkeley Books, 1999).

—, *The Search: More of the Ancient Art of the New Survival* (New York: Berkeley Books, 1980).

Bryden, H. Anderson, *Gun and Camera in Southern Africa* (London: Edward Stanford, 1893).

Buijtenhuijs, R., *Mau Mau: Twenty Years After: The Myth and the Survivors* (New York: Mouton, 1973).

Burbridge, B., *Tracking and Capturing the Ape-Man of Africa* (New York: The Century Company, 1928).

Burnham, F. R., *Scouting on Two Continents* (Garden City, NY: Garden City Publishing, 1926).

Bush, P. and P. Thomas, *Peter Bush: A Life in Focus* (Auckland: Hodder Moa, 2009).

Callwell, C. E., *Small Wars: Their Principles and Practise* (1896; Lincoln, NE: University of Nebraska Press, 1996).

Campbell, G., *The Charging Buffalo: A History of the Kenya Regiment, 1937–1963* London: Leo Cooper, 1986.

Campling, A. B., Major General, DCD, 'Pseudo-Terrorist Operations in Rhodesia' (2006).

Cann, J. P. *Counterinsurgency in Africa: The Portuguese Way of War, 1961–1974* (Solihull, UK: Helion and Company, 2012).

—, *The Flechas: Insurgent Hunting in Eastern Angola, 1965–1974* (Pinetown, South Africa: 30 Degrees South, 2013).

Capstick, P. Hathaway, *Warrior: The Legend of Colonel Richard Meinertzhagen* (New York: St Martin's Press, 1998).

—, *Death in the Long Grass: A Big Game Hunter's Adventures in the African Bush* (New York: St Martin's Press, 1977).

Carruthers, P., 'The Roots of Scientific Reasoning: Modularity and the Art of Tracking', in P. Carruthers, S. Stich and M. Siegal (eds), *The Cognitive Basis of Science* (Cambridge: Cambridge University Press, 2002), pp. 73–96.

Carss, B., *The SAS Guide to Tracking* (New York: Lyons Press, 1999).

Castle, I., *The British Infantryman in South Africa, 1877–81* (Oxford: Osprey, 2003).

Cawthra, G., *Brutal Force: The Apartheid War Machine* (London: International Defence and Aid Fund, 1986).

Chappell, S., 'Air Power in the Mau Mau Conflict: The Government's Chief Weapon', *The RUSI Journal*, 156:1 (February-March 2011), pp. 64–70.

Chilvers, B., 'Rhino's Last Stand in Africa', *REF Journal*, 3 (1990), p. 17.

Christy, C., 'The Ituri River, Forest and Pygmies', *The Geographic Journal*, 46:3 (September 1915), pp. 205–6.

Cilliers, J. K., *Counter-Insurgency in Rhodesia* (Beckenham: Croom Helm Publishing, 1985).

Cipolletta, C., 'The Long Road to Habituation: A Window into the Lives of Gorillas', in M. M. Robbins and C. Boesch (eds), *Among African Apes: Stories from the Field* (Berkeley, CA: University of California Press, 2011), pp. 129–42.

Clough, M., *Mau Mau Memoirs: History, Memory and Politics* (Boulder, CO: Lynne Rienner Publishers, 1998).

Cock, J. and L. Bernstein, *Melting Pots and Rainbow Nations: Conversations about Difference in the United States and South Africa* (Chicago, IL: University of Illinois Press, 2002).

Cole, B., *The Elite: The Story of the Rhodesian Special Air Service* (Amazimtoti, South Africa: Three Knights Publishing, 1984).

Comber, L., *Malaya's Secret Police, 1945–60: The Role of the Special Branch in the Malayan Emergency* (Victoria, Australia: Monash University Press, 2008).

Corfield, F. D., *The Origins and Growth of Mau Mau: An Historical Survey* (Nairobi: Colony and Protectorate of Kenya, 1960).

Correia, P. E. S., 'Political Relations Between Portugal and South Africa from the End of the Second World War until 1974' (PhD Thesis, University of Johannesburg, 2007).

Cross, J., *Jungle Warfare; Experiences and Encounters* (Barnsley: Pen and Sword, 2008).

Croukamp, D., *The Bush War in Rhodesia: The Extraordinary Combat Memoir of a Rhodesian Reconnaissance Specialist* (Boulder, CO: Paladin Press, 2006).

de Kock, E., *A Long Night's Damage: Working for the Apartheid State* (Saxonwold: Contra Press, 1998).

de Vries, J. and S. Swart, 'The South African Defence Force and Horse Mounted Infantry Operations, 1974–1985', *Scientia Militaria*, 40:3 (2012), pp. 398–428.

Desmond Clark, J., 'Bushmen Hunters of the Barotse Forests', *Northern Rhodesia Journal*, 1:3 (1951), pp. 56–65.

Diaz, D. with V. L. McCann, *Tracking: Signs of Man, Signs of Hope: A Systematic Approach to the Art and Science of Tracking Humans* (Guilford, CT: Lyons Press, 2005).

Dieckmann, U., *Haillom in the Etosha Region: A History of Colonial Settlement, Ethnicity and Nature Conservation* (Basel, Switzerland: Basler Afrika, 2007).

Dippenaar, M., *The History of the South African Police, 1912–1988* (Pretoria: Promedia, 1988).

Doke, C. M., *The Lambas of Northern Rhodesia: A Study of Their Customs and Beliefs* (London: George G. Harrap, 1931).

Donelan, D. S., *Tactical Tracking Operations: The Essential Guide for Military and Police Trackers* (Boulder, CO: Paladin Press, 1998).

Dornan, S. S., 'The Tati Bushmen (Masarwas) and their Language', *The Journal of the Royal Anthropological Institute of Great Britain and Ireland*, 47 (January–June 1917), pp. 37–112.

Dreschler, H., *Let Us Die Fighting: The Struggle of the Herero and Nama Against German Imperialism (1884–1915)* (London: Zed Press, 1980).

Du Toit, B., *The Boers in East Africa: Ethnicity and Identity* (Westport, CT: Greenwood, 1998).

Duffy, R., 'The Role and Limitations of State Coercion: Anti-Poaching Policies in Zimbabwe', *Journal of Contemporary African Studies*, 17:1 (1999), pp. 97–120.

Dunlay, T. W., *Wolves for the Blue Soldiers: Indian Scouts and Auxiliaries with the United States Army* (Lincoln, NE: University of Nebraska Press, 1982).

Durand, A., *Zulu Zulu Foxtrot: To Hell and Back with Koevoet* (Cape Town: Zebra Press, 2012).

—, *Zulu Zulu Golf: Life and Death with Koevoet* (Cape Town: Zebra Press, 2011).

Ekandjo, P., *The Jungle Fighter* (Windhoek: privately published, 2011).

—, *The Volunteers' Army, 1962–1989* (Windhoek: privately published, 2014).

Elkins, C., *Imperial Reckoning: The Untold Story of Britain's Gulag in Kenya* (New York: Henry Holt, 2005).

Ellert, H., *Rhodesian Front War: Counter-Insurgency and Guerrilla War in Rhodesia, 1962–80* (Gweru: Mambo Press, 1989).

Ellis, S., 'Of Elephants and Men: Politics and Nature Conservation in South Africa', *Journal of Southern African Studies*, 20:1 (March 1994), pp. 53–69.

—, *External Mission: The ANC in Exile, 1960–90* (Oxford: Oxford University Press, 2013).

Els, P. J., *We Fear Naught But God* (n.p.: privately published, 2009).

Essex-Clark, J., *Maverick Soldier: An Infantryman's Story* (Melbourne: Melbourne University Press, 1991).

Esterhuyse, A. and E. Jordaan, 'The South African Defence Force and Counterinsurgency, 1966–1990', in D.-P. Baker and E. Jordaan (eds), *South Africa and Contemporary Counterinsurgency: Roots, Practises and Prospects* (Claremount, South Africa: International Publishers, 2010), pp. 104–23.

Fey, V., *Cloud Over Kenya* (London: Collins, 1964).

—, *Wide Horizon: Tales of a Kenya That Has Passed into History* (New York: Vantage Press, 1982).

Foran, W. R., *Kill or be Killed: The Rambling Reminiscences of an Amateur Hunter* (London: Hutchinson, 1933).

Franklin, D., *A Pied Cloak: Memoirs of a Colonial Police (Special Branch) Officer* (London: Janus Publishing, 1996).

French, P., *To the Edge: Shadows of a Forgotten Past with the Rhodesian SAS and Selous Scouts* (Solihull, UK: Helion and Company, 2012).

Frump, R., *The Man-Eaters of Eden: Life and Death in Kruger National Park* (New York: Lyons Press, 2006).

Furedi, F., *The Mau Mau War in Perspective* (Oxford: James Currey, 1989).

Gatti, A., *Great Mother Forest* (New York: Charles Scribner's Sons, 1937).

Geldenhuys, J., *At the Front: A General's Account of South Africa's Border War* (Johannesburg: Jonathan Balls, 2009).

George, E., *The Cuban Intervention in Angola, 1965–1991: From Che Guevara to Cuito Cuanavale* (London: Frank Cass, 2005).

Gewald, J.-B., *Herero Heroes: A Socio-Political History of the Herero of Namibia, 1890–1923* (Oxford: James Currey, 1999).

Gibbs, P., H. Phillips and N. Russell, *Blue and Old Gold: The History of the British South Africa Police, 1889–1980* (Johannesburg: 30 Degrees South, 2009).

Gill, L., *Remembering the Regiment* (Victoria, British Columbia: Trafford, 2004).

Gillmore, P., *The Great Thirst Land: A Ride Through Natal, Orange Free State and the Kalahari Desert* (London: Cassell, Peter and Galpin, 1878).

—, *The Hunter's Arcadia* (London: Chapman and Hall, 1886).

Gleijeses, P., *Conflicting Missions: Havana, Washington and Africa, 1959–1976* (Chapel Hill, NC: University of North Carolina Press, 2002).

Goodrich, T., *Black Flag: Guerilla Warfare on the Western Border, 1861–65* (Bloomington, IN: Indiana University Press, 1999).

Gordon, R. and S. Shotlo Douglas, *The Bushman Myth: The Making of a Namibian Underclass* (Boulder, CO: Westview Press, 2000).

Gordon, R. J., '"Captured on Film": Bushmen and the Claptrap of Performative Primitives', in P. Landou and D. Kaspin (eds), *Images and Empire: Visuality in Colonial and Post-Colonial Africa* (Berkeley, CA: University of California Press, 2002), pp. 212–32.

Gordon, R., 'People of the Great White Lie?', *Cultural Survival Quarterly*, 15:1 (Spring 1991), at https://www.culturalsurvival.org/ourpublications/csq/article/people-great-white-lie [accessed 18 April 2015].

Gordon-Cumming, R., *Five Years of a Hunter's Life in the Far Interior of South Africa*, 2 vols (London: John Murray, 1850).

Greeff, J., *A Greater Share of Honour: The Memoirs of a Recce Officer* (Durban: Just Done, 2008).

Grob-Fitzgibbon, B., *Imperial Endgame: Britain's Dirty Wars and the End of Empire* (London: Palgrave-MacMillan, 2011).

Grossman, A., 'Lost COIN: The Erosion of Counterinsurgency in the South African Army', in Baker and Jordaan, *South Africa and Contemporary Counterinsurgency*, pp. 136–51.

Grundlingh, A., '"Protectors and Friends of the People?" The South African Constabulary in the Transvaal and Orange River Colony, 1900–1908', in D. M. Anderson and D. Killingray (eds), *Policing the Empire: Government, Authority and Control, 1830–1940* (Manchester: Manchester University Press, 1991), pp. 168–82.

Grundy, K., *Soldiers Without Politics: Blacks in the South African Armed Forces* (Berkeley, CA: University of California Press, 1983).

Grundy, T. and B. Miller, *The Farmer at War* (Salisbury: Modern Farming Publications, 1979).

Hallam Parr, Sir H., *A Sketch of the Kafir and Zulu Wars: From Guadana to Isandhlwana* (London: C. Kegan Paul, 1880).

Hamann, H., *Days of the Generals: The Untold Story of South Africa's Apartheid Era Military Generals* (Cape Town: Zebra Press, 2001).

Harrison, S., *Dark Trophies: Hunting and the Enemy Body in Modern War* (New York: Berghahn Books, 2012).

Haydon, A. L. *The Trooper Police of Australia: A Record of Mounted Police Work in the Commonwealth from the Earliest Days of Settlement to the Present Time* (London: Andrew Melrose, 1911).

Heather, R. W., 'Of Men and Plans: the Kenya Campaign as Part of the British Counter-Insurgency Experience', *Conflict Quarterly* (Winter 1993), pp. 17–26.

Heitman, H. R., *South African War Machine* (Novato, CA: Presidio Press, 1985).

—, *South African Armed Forces* (Cape Town: Buffalo Productions, 1990).

Herbstein, D. and J. Evenson, *The Devils Are Among Us: The War for Namibia* (London: Zed Press, 1989).

Hespeler-Boultbee, J. J. *Mrs. Queen's Chump: Idi Amin, Communists and Other Silly Follies of the British Empire: A Military Memoir* (British Columbia, Canada: CCB Publishing, 2012).

Hewitt, P., *Kenya Cowboy: A Police Officer's Account of the Mau Mau Emergency* (Johannesburg: 30 Degrees South, 2008).

Hiscox, C. L., *The Dawn Stand-to: The Life of IVB (Peter) Mills, QPM, CPM* (Bideford, UK: Edward Gaskell Publishers, 2000).

Hitchcock, R. K., 'Centralization, Resource Depletion and Coercive Conservation Among the Tyua of the Northeastern Kalahari', *Human Ecology*, 23:2 (June 1995), pp. 169–98.

—, 'Refugees, Resettlement, and Land and Resource Conflicts: The Politics of Identity Among !Xun and Khwe San in Northeastern Namibia', *African Study Monographs*, 33:2 (June 2012), pp. 73–132.

—, M. Sapignoli and W. A. Babchuk, 'Settler Colonialism, Conflicts, and Genocide: Interactions Between Hunter-Gatherers and Settlers in Kenya, and Zimbabwe and Northern Botswana', *Settler Colonial Studies*, 5:1 (2015), pp. 40–65.

Hoffman, M. T., 'Major P. J. Pretorius and the Decimation of the Addo Elephant Herd in 1919–20: Important Reassessments', *Koedoe*, 36:2 (1993), pp. 23–44.

Holman, D., *Elephants at Sundown: The Story of Bill Woodley* (London: W. H. Allen, 1978).

—, *The Elephant People* (London: James Murray, 1967).

Holub, E., *Seven Years in South Africa: Travels, Researches and Hunting Adventures Between the Diamond Fields and the Zambezi, 1872–79*, 2 vols (London: Sampson Low, 1881), vol. 1.

Hooper, J., *Koevoet: Experiencing South Africa's Deadly Bush War* (1988; Warwickshire: GG Books, 2012).

Hopkins, C., 'Tarzan's Africa: Living a Boyhood Dream of Africa Inspired by Johnny Weissmuller, the Author Hunts the African Forest with Pygmy Trackers', *Wild Hunting*, 7 (2006), at www.surefire.com [accessed 13 April 2015].

Hornsby, C., *Kenya: A History Since Independence* (London: I. B. Tauris, 2013).

House, A., *The Great Safari: The Lives of George and Joy Adamson* (New York: William Morrow, 1993).

Hubbard, D. H. Jr, *Bound for Africa: Cold War Fight Along the Zambezi* (Annapolis, MD: Naval Institute Press, 2008).

Huggonson, D., 'The Black Trackers of Bloemfontein', *Land Rights News*, February 1990, p. 20.

Hughes, D. McDermott, 'Whites Lost and Found: Immigration and Imagination in Savanna Africa', in B. Caminero-Santangelo and G. Myers (eds), *Environment at the Margins: Literary and Environmental Studies in Africa* (Athens, OH: Ohio University Press, 2011).

Hummel, C. (ed.), *The Frontier War Journal of Major John Crealock 1878* (Cape Town: Van Riebeeck Society, 1988).

Hunter, J. A., *Hunter's Tracks* (London: Hamish Hamilton, 1957).

Hunter, M. L, R. K. Hitchcock and B. Wyckoff-Baird, 'Women and Wildlife in Southern Africa', *Conservation Biology*, 4:4 (December 1990), pp. 448–51.

Hurth, J. D. and J. W. Brokaw, 'Visual Tracking and the Military Tracking Team Capability: A Disappearing Skill and Misunderstood Capability', *Small Wars Journal* (2010), pp. 1–17.

Hurth, J., *Combat Tracking Guide* (Mechanicsburg, PA: Stackpole Books, 2012).

Huxley, E., *Nine Faces of Kenya* (London: Harvill Press, 1991).

—, *The Mottled Lizard* (London: Chatto and Windus, 1962).

Inscoe, J. C. and G. B. McKinney, *The Heart of Confederate Appalachia: Western North Carolina in the Civil War* (Chapel Hill, NC: University of North Carolina Press, 2000).

Ionides, C. J. P., *Mambas and Man-Eaters: A Hunter's Story* (New York: Holt, Rinehart and Winston, 1966).

Itote, W. (General China), *Mau Mau General* (Nairobi: East African Publishing House, 1967).

Jaster, R. S., 'South African Defense Strategy and the Growing Influence of the Military', in W. Foltz and H. Bienen (eds), *Arms and the African: Military Influences on Africa's International Relations* (New Haven, CT: Yale University Press, 1985).

Jeal, T., *Baden-Powell: Founder of the Boy Scouts* (New Haven, CT: Yale University Press, 2007).

Jones, S., *Under the African Sun* (London: Hurst and Brackett, 1956).

—, *From Boer War to World War: Tactical Reform of the British Army, 1902–1914* (Norman, OK: University of Oklahoma Press, 2012).

Kamongo, S. and L. Bezuidenhout, *Shadows in the Sand: A Koevoet Tracker's Story of an Insurgency War* (Johannesburg: 30 Degrees South Publishers, 2011).

Katjavivi, P. H., *A History of Resistance in Namibia* (London: James Currey, 1988).

Kerwin, D., 'The Lost Trackers: Aboriginal Servicemen in the 2nd Boer War', *Sabretache*, 54:1 (March 2013), p. 4–14.

Kitson, F., *Bunch of Five* (London: Faber and Faber, 1977).

—, *Gangs and Counter-Gangs* (London: Barrie, 1960).

—, *Low Intensity Operations: Subversion, Insurgency and Peacekeeping* (London: Faber and Faber, 1971).

Kolata, G. B., '!Kung Bushmen Join the South African Army', *Science*, 211:4482 (February 1981), pp. 562–4.

Kondlo, K. M., '"In the Twilight of the Azanian Revolution": The Exile History of the Pan Africanist Congress of Azania (South Africa) (1960–1990)' (PhD Thesis, Rand Afrikaans University, 2003).

Korff, G., *19 With a Bullet: A South African Paratrooper in Angola* (Johannesburg: 30 Degrees South, 2009).

Krantz, G., 'Brain Size and Hunting Ability in Earliest Man', *Current Anthropology*, 9:5 (December 1968), pp. 450–1.

le Cordeur, B. (ed.), *The Journal of Charles Lennox Stretch* (Grahamstown: Longman, 1988).

Leakey, L. S. B., *Mau Mau and the Kikuyu* (London: Routledge, 1952).

Leakey, R., 'Early Humans: Of Whom Do We Speak?' in F. E. Grine, J. G. Fleagle and R. E. Leakey (eds), *The First Humans: Origin and Evolution of the Genus Homo* (London: Springer, 2009), pp. 3-6.

Lee, R. and S. Hurlich, 'From Foragers to Fighters: South Africa's Militarization of the Namibian San', in E. Leacock and R. Lee (eds), *Politics and History in Band Societies* (Cambridge: Cambridge University Press, 1982), pp. 327–46.

Lee, R. B., *The !Kung San: Men, Women, and Work in a Foraging Society* (Cambridge: Cambridge University Press, 1979).

Lemon, D., *Never Quite a Soldier: A Rhodesian Policeman's War 1971–82* (Stroud: Albita Books, 2000).

Lewis, J., *The Batwa Pygmies of the Great Lakes Region* (London: Minority Rights Group International, 2000).

Leys, C. and J. Saul, *Namibia's Liberation Struggle: The Two Edged Sword* (London: James Currey, 1995).

Liebenberg, L., 'Persistence Hunting by Modern Hunter-Gatherers', *Current Anthropology*, 47:6 (December 2006), pp. 1017–26.

—, *A Photographic Guide to Tracks and Tracking in Southern Africa* (Cape Town: Struik, 2000).

—, *The Art of Tracking: The Origin of Science* (Claremont, South Africa: David Phillip, 1990).

—, A. Louw and M. Elbroch: *Practical Tracking: A Guide to Following Footprints and Finding Animals* (Mechanicsburg, PA: Stackpole Books, 2010).

Lineberry, W. P., *East Africa* (New York: W. H. Wilson, 1968).

Lord, D., *From Fledgling to Eagle: The South African Air Force During the Border War* (Johannesburg: 30 Degrees South Publishers, 2008).

Lott, J., 'Run the Bastards Down: C.A.T.U. Tracks Terrorists; Rhodesia`s Civilian Tracking Unit', *Soldier of Fortune Magazine*, July 1979, pp. 46–51.

MacDonald, R., *Sons of Empire: The Frontier and the Boy Scout Movement, 1890–1918* (Toronto: University of Toronto Press, 1993).

MacKenzie, J. M., *The Empire of Nature: Hunting, Conservation and British Imperialism* (New York: St Martin's Press, 1988).

MacMillan, H., *The Lusaka Years: The ANC in Exile in Zambia, 1963–1994* (Auckland Park, South Africa: Jacana Media, 2013).

Main, M., 'Notes on Bushman Tracking Techniques', unpublished paper, 22 May 2009.

Mamdani, M., *Citizen and Subject: Contemporary Africa and the Legacy of Late Colonialism* (Princeton, NJ: Princeton University Press, 1996).

Mandiringana, E. and T. Stapleton, 'The Literary Legacy of Frederick Courteney Selous', *History in Africa*, 25 (1998), pp. 199–218.

Mangan, J. A. and C. McKenzie, *Militarism, Hunting, Imperialism: "Blooding" the Martial Male* (New York: Routledge, 2010).

Marshall Thomas, E., *The Old Way: A Story of the First People* (New York: Picador, 2006).

Marshall, L., 'Marriage Among !Kung Bushmen', *Africa: Journal of the International Africa Institute*, 29:4 (October 1959), pp. 335–65.

Martin, D. and P. Johnson, *The Struggle for Zimbabwe: The Chimurenga War* (New York: Monthly Review, 1981).

McAleese, P., *No Mean Soldier: The Story of the Ultimate Professional Soldier in the SAS and Other Forces* (London: Orion, 1994).

McCallion, H. (Henry Gow), *Killing Zone: A Life in the Paras, the Recces, the SAS and the RUC* (London: Bloomsbury, 1996).

McCuen, J. J., *The Art of Counter-Revolutionary War: The Strategy of Counter-Insurgency* (Harrisberg: Stackpole Books, 1966).

McGibbon, I., *New Zealand's Vietnam War: A History of Combat, Commitment and Controversy* (Auckland: Exile Publishing, 2010).

McMullin, J., *Ex-Combatants and the Post-Conflict State: Challenges of Reintegration* (London: Palgrave-MacMillan, 2013).

Meadows, K., *Sometimes When It Rains: White Africans in Black Africa* (Portland, OR: Thorntree Press, 2000).

Mears, R., *My Outdoor Life* (London: Hodder and Stoughton, 2013).

Merritt, S., *Seek On* (Houston, TX: Strategic Book Publishing, 2011).

Miller, C., *Painting the Map Red: Canada and the South African War* (Toronto: McGill-Queen's Press, 1993).

Milton, J., *The Edges of War: A History of Frontier Wars (1702–1878)* (Cape Town: Juta, 1983).

Minter, W., *Apartheid's Contras: An Inquiry into the Roots of War in Angola and Mozambique* (London: Zed Books, 1994).

Moodie, D. C. F., *The History of the Battles and Adventures of the British, the Boers and the Zulus in Southern Africa* (Sydney, Australia: George Robertson, 1879).

Moorcroft, P. and P. McLaughlin, *The Rhodesian War: A Military History* (London: Pen and Sword, 2008).

Moreman, T., *The Jungle, the Japanese and the British Commonwealth Armies at War 1941–45: Fighting Methods, Doctrine and Training for Jungle Warfare* (New York: Frank Cass, 2005).

Mostert, N., *Frontiers: The Epic of South Africa's Creation and the Tragedy of the Xhosa People* (New York: Alfred Knopf, 1992).

Mumford, A., *The Counter-Insurgency Myth: The British Experience of Irregular Warfare* (New York: Routledge, 2012).

Nagl, J. A., *Learning to Eat Soup with a Knife: Counter-Insurgency Lessons From Malaya and Vietnam* (Chicago, IL: University of Chicago Press, 2002).

Namakalu, O. O., *Armed Liberation Struggle: Some Accounts of PLAN's Combat Operations* (Windhoek, Namibia: Gamsberg MacMillan, 2004).

Nasson, B., *Abraham Esau's War: A Black South African's War in the Cape, 1899–1902* (Cambridge: Cambridge University Press, 1991).

—, *Springboks on the Somme: South Africa and the Great War, 1914–1918* (Johannesburg: Penguin, 2007).

Nhongo-Simbanegavi, J., *For Better or Worse: Women and ZANLA in Zimbabwe's Liberation Struggle* (Harare: Weaver Press, 2000).

Nielsen, S., "The White Devil of Mozambique: One Man Army Against FRELIMO," *Soldier of Fortune* (October 1978), pp. 78–83.

Nortje, P., *32 Battalion: The Inside Story of South Africa's Elite Fighting Unit* (Cape Town: Zebra Press, 2003).

Norval, M., 'SADF's Bushman Battalion: Desert Nomads on SWAPO's Track', *Soldier of Fortune* (March 1984), pp. 71–5.

Norval, M., *Death in the Desert: The Namibian Tragedy* (Washington D.C.: Selous Foundation, 1989).

Nusselhuf, F. J., 'General Paul Von Lettow-Vorbeck's East Africa Campaign: Maneuver Warfare on the Serengeti' (MA Thesis, University of North Texas, 2012).

Orford, B., *Kamchacha: Rhodesian Game Ranger* (Bulawayo, Zimbabwe: privately published, 2008).

Page, M., *KAR: A History of the King's African Rifles* (London: Leo Cooper, 1998).

Paice, E., *Tip and Run: The Untold Tragedy of the Great War in Africa* London: Weidenfeld and Nicolson, 2007.

Parker, I., *The Last Colonial Regiment: The History of the Kenya Regiment (TF)* (Moray: Librario Publishing, 2009).

Parker, J., *Assignment Selous Scouts: Inside Story of a Rhodesian Special Branch Officer* (Alberton, South Africa: Galago, 1999).

Parsons, T., *Race, Resistance and the Boy Scout Movement in British Colonial Africa* (Athens, OH: Ohio University Press, 2004).

—, *The African Rank-and-File: Social Implications of Colonial Military Service in the King's African Rifles, 1902–1964* (Oxford: James Currey, 1999).

Peled, A., *A Question of Loyalty: Military Manpower Policy in Multiethnic States* (Ithaca, NY and London: Cornell University Press, 1999).

Peluso, N. L. and P. Vandergeest, 'Political Ecologies of War and Forests: Counter-Insurgencies and the Making of National Natures', *Annals of the Association of American Geographers*, 101:3 (2011), pp. 587–608.

Percival, A. B., *A Game Ranger's Notebook* (London: Nisbet, 1924).

Peterson, J., *Oman's Insurgencies: The Sultanate's Struggle for Supremacy* (London: Saqi, 2007).

Petter-Bowyer, P. J. H., *Winds of Destruction: The Autobiography of a Rhodesian Combat Pilot* (Johannesburg: 30 Degrees South, 2005).

Pickering, T. R., *Rough and Tumble: Aggression, Hunting and Human Evolution* (Berkeley, CA: University of California Press, 2013).

Pittaway, J. and C. Fourie (eds), *SAS Rhodesia* (Musgrave, South Africa: Dandy Agencies, 2003).

Player, I., *Zulu Wilderness: Shadow and Soul* (Golden, CO: Fulcrum Publishing, 1998).

Pocock, G. A., *Outrider of Empire: The Life and Adventures of Roger Pocock* (Edmonton, Alberta: University of Alberta Press, 2007).

Polack, P., *The Last Hot Battle of the Cold War: South Africa vs. Cuba in the Angolan Civil War* (Oxford: Casemate, 2013).

Porch, D., *Counter-Insurgency: Exposing the Myths of the New Way of War* (Cambridge: Cambridge University Press, 2013).

Powell-Cotton, P. H. G., 'A Journey Through the Eastern Portion of the Congo State', *The Geographic Journal*, 30:4 (October 1907), pp. 371–82.

—, 'Notes on a Journal Through the Great Ituri Forest', *Journal of the Royal African Society*, 7:25 (October 1907), pp. 1–12.

Pretorius, P. J., *Jungle Man: The Autobiography of Major P. J. Pretorius* (1947; Alexander, NC: Alexander Books, 2001).

Pugsley, C., *From Emergency to Confrontation: The New Zealand Armed Forces in Malaya and Borneo, 1949–66* (Oxford: Oxford University Press, 2003).

Quan, H., *Sam Steele: The Wild West Adventures of Canada's Most Famous Mountie* (Canmore, Canada: Altitude Publishing, 2003).

Rainsford, W. S., *The Land of the Lion* (New York: Doubleday, Page and Coy, 1909).

Ralinala, R. M., J. Sithole, G. Houston and B. Magubane, 'The Wankie and Sipolilo Campaigns', in South African Democracy Education Trust, *The Road to Democracy in South Africa Vol. 1 (1960–70)* (Cape Town: Zebra Press, 2004), pp. 479–540.

Ranger, T. and N. Bhebhe (eds), *Soldiers in Zimbabwe's Liberation War* (London: James Currey, 1995).

Ranger, T., *Peasant Consciousness and Guerrilla War in Zimbabwe: A Comparative Study* (London: James Currey, 1985).

Reeve, R. and S. Ellis, 'An Insider's Account of the South African Security Forces' Role in the Ivory Trade', *Journal of Contemporary African Studies*, 13:2 (1995), pp. 227–44.

Reid Daly, R. (as told to P. Stiff), *Selous Scouts: Top Secret War* Alberton: Galago, 1982.

—, "War in Rhodesia: Cross-Border Operations," in A.J. Venter (ed.), *Challenge: Southern Africa Within the African Revolutionary Context* (Gibraltar: Ashanti Publishing, 1989), pp. 146–182.

Reitz, D., *Command: A Boer Journal of the Boer War* (London: Faber and Faber, 1929).

Reuning, H. and W. Wortley, 'Psychological Studies of the Bushman', *Psychologica Africana*, 7 (1973), pp. 1–113.

Richards, J., *The Secret War: A True History of Queensland's Native Police* (St Lucia: University of Queensland Press, 2008).

Rico, M., *Nature's Noblemen: Trans-Atlantic Masculinities and the Nineteenth Century American West* (Princeton, NJ: Yale University Press, 2013).

Robbins, D., *On the Bridge of Goodbye: The Story of South Africa's Discarded San Soldiers* (Johannesburg: Jonathan Ball Publishers, 2007).

Robins, S. and K. Van Der Waal, 'Model Tribes and Iconic Conservationists? Tracking the Makuleke Restitution Case in Kruger National Park', in C. Walker, A. Bohlin, R. Hall and T. Kepe (eds), *Land, Memory, Reconstruction, and Justice: Perspectives on Land Claims in South Africa* (Athens, OH: Ohio University Press, 2010), pp. 163–80.

Robinson, P., 'The Search for Mobility During the Second Boer War', *Journal of the Society for Army Historical Research*, 68:346 (2008), pp. 140–57.

Roosevelt, T., *African Game Trails: An Account of the African Wanderings of an American Hunter-Naturalist* (New York: Charles Scribner and Sons, 1921).

Rosberg, C. G. and J. C. Nottingham, *The Myth of "Mau Mau:" Nationalism in Kenya* (New York: Praeger, 1966).

Sarkin, J., *Germany's Genocide of the Herero: Kaiser Wilhelm II, His General, His Settlers, His Soldiers* (Woodbridge, UK: James Currey, 2011).

Sarnecki, J., 'The Emergence of Empathy in Cross-species Mind Reading', in L. Swan (ed.), *Origins of Mind* (London: Springer, 2013).

Sass, B., 'The Union and South African Defence Force, 1912 to 1994', in J. Cilliers and M. Reichardt (eds) *About Turn: The Transformation of the South African Military and Intelligence* (Halfway House: IDP, 1995), pp. 118–39.

Schafer, D., *Simply the Greatest Life: Finding Myself in the Country* (Bloomington, IN: Balboa Books, 2012).

Scholtz, L., *The SADF in the Border War, 1966–1989* (Cape Town: Tafelberg, 2013).

Scott-Donelan, D., 'Zambezi Valley Manhunt', *Soldier of Fortune*, March 1985.

Seegers, A., 'One State, Three Faces; Policing in South Africa (1910–1990)', *Social Dynamics*, 17:1 (1991), pp. 36–48.

—, *The Military in the Making of Modern South Africa* (London: I. B. Tauris, 1996).

Selley, R., *West of the Moon: Early Zululand and a Game Ranger at War in Rhodesia* (Johannesburg: 30 Degrees South, 2009).

Selous, F. C., *Sunshine and Storm in Rhodesia* (London: Rowland Ward, 1896).

—, *Travel and Adventure in South Eastern Africa* (London: Rowland Ward, 1893).

Shear, K., 'Police Dogs and State Rationality in Early Twentieth Century South Africa', in L. Van Sittert and S. Swart (eds), *Canis Africanis: A Dog History of Southern Africa* (Leiden: Brill, 2008), pp. 193–216.

Sheldrick, D., *An African Love Story: Love, Life and Elephants* (London: Penguin, 2012).

Sibanda, E., *The Zimbabwe African People's Union, 1961–1987* (Trenton: Africa World Press, 2005).

Sinclair, G., *At the End of the Line: Colonial Policing and the Imperial Endgame* (Manchester: Manchester University Press, 2006).

Smiley, D. and P. Kemp, *Arabian Assignment* (London: Leo Cooper, 1975).

Speed, N. G., *Born to Fight: Major Charles Joseph Ross, A Definitive Study of His Life* (Melbourne: The Caps and Flints Press, 2002).

Stapleton, T., *Faku: Rulership and Colonialism in the Mpondo Kingdom, c.1780–1867* (Waterloo: Wilfrid Laurier University Press, 2001).

Stark, P., *The White Bushman* (Pretoria: Protea Book House, 2011).

Steele, N., *Bushlife of a Game Warden* (Cape Town: T. V. Bulpin, 1979).

Steenkamp, W., 'Politics of Power – The Border War', in A. J. Venter (ed.), *Challenge: Southern African Within the African Revolutionary Context – An Overview* (Gibraltar: Ashanti Publishing, 1989), pp. 207–23.

—, *Borderstrike! South Africa into Angola, 1975–80* (Durban: Just Done Productions, 2006).

—, *South Africa's Border War, 1966–1989* (Gibraltar: Ashanti, 1989).

Steinhart, E., *Black Poachers, White Hunters: A Social History of Hunting in Colonial Kenya* (Oxford: James Currey, 2006).

Stevenson-Hamilton, J., *The Low-Veld: Its Wild Life and Its People* (London: Cassell and Company, 1927).

Stewart, R., *Sam Steele: Lion of the Frontier* (Toronto: Doubleday, 1979).

Stiff, P., *The Covert War:Koevoet Operations Namibia, 1979–89* (Alberton, South Africa: Galago, 2004).

—, *The Silent War; South Africa Recce Operations, 1969–1994* (Alberton, South Africa: Galago, 2006).

Stigand, C. H., *The Game of British East Africa* (London: Harper Cox, 1909).

Storey, W., *Guns, Race and Power in Colonial South Africa* (Cambridge: Cambridge University Press, 2008).

Strachan, H., *The First World War in Africa* (Oxford: Oxford University Press, 2004).

Tatham, G. H., 'The Rhino Conservation Strategy in the Zambezi Valley Codenamed Operation Stronghold', *The Zimbabwe Science News*, 22:1–2 (January–February 1988).

Taylor, J., *Pondoro: The Last of the Ivory Hunters* (New York: Simon and Schuster, 1955).

Taylor, S., *The Mighty Nimrod: A Life of Frederick Courteney Selous, African Hunter and Adventurer, 1851–1917* (London: Harper Collins, 1989).

Thomas, K., *Shadows in an African Twilight* (Cape Town: Uthekwane Press, 2008).

Thompson, J. H., *An Unpopular War: From Afkak to Bosbefok: Voices of South African National Servicemen* (Cape Town: Zebra Press, 2006).

Thompson, P., *Black Soldiers of the Queen: The Natal Native Contingent in the Anglo-Zulu War* (Tuscaloosa, AL: University of Alabama Press, 2006).

Thompson, R. G. K., *Defeating Communist Insurgency: Experiences from Malaya and Vietnam* (London: Chatto and Windus, 1966).

Throup, D., *Economic and Social Origins of Mau Mau, 1945–53* (Oxford: James Currey, 1987).

Tobias, P. V. and M. Biesele, *The Bushmen: San Hunters and Herders of Southern Africa* (Cape Town: Human and Rousseau, 1978).

Tonsetic, R., *Warriors: An Infantryman's Memoir of Vietnam* (Novato, CA: Presidio Press, 2004).

Tredger, N., *From Rhodesia to Mugabe's Zimbabwe: Chronicles of a Game Ranger* (Alberton, South Africa: Galago, 1999).

Trethowan, A., *Delta Scout: Ground Coverage Operator* (Johannesburg: 30 Degrees South, 2008).

Turner, J. W., *Continent Ablaze: The Insurgency Wars in Africa 1960 to the Present* (Johannesburg: Jonathan Ball, 1998).

Udogu, E. I., *Liberation Namibia: The Long Diplomatic Struggle Between the United Nations and South Africa* (Jefferson, CO: Macfarland, 2012).

Uys, I., *Bushman Soldiers: Their Alpha and Omega* (Germiston, South Africa: Fortress Publications, 1993).

Vail, L. (ed.), *The Creation of Tribalism in Southern Africa* (Berkeley, CA: University of California Press, 1991).

Van Der Waals, W. S., *Portugal's War in Angola, 1961–1974* (Pretoria: Protea Book House, 2011).

van Riel, F., *My Life with Leopards: Graham Cooke's Story* (London: Penguin Books, 2013).

Van Vuuren, L., "'And He Said They Were Ju/Wasi, the People...': History and Myth in John Marshall's "Bushman Films" 1957–2000', *Journal of Southern African Studies*, 35:3 (2009), pp. 557–74.

—, 'The Many Myths of Laurens van der Post: Van Der Post and Bushmen in the Television Series Lost World of Kalahari (1958)', *South African Historical Journal*, 48:1 (2003), pp. 47–60

Venning, J. H., 'A Lesson in Lion Hunting', *Northern Rhodesia Journal*, 3:5 (1958), pp. 425–7.

Venter, A. J., *Gunship Ace: The Wars of Neall Ellis, Helicopter Pilot and Mercenary* (Newbury, UK: Casemate, 2011).

— and friends, *War Stories: Up Close and Personal in Third World Conflicts* (Pretoria: Protea Book House, 2011).

—, *Portugal's Guerrilla Wars in Africa: Lisbon's Three Wars in Angola, Mozambique and Portuguese Guinea 1961–74* (Solihull, UK: Helion and Company, 2013).

Vines, A., *Still Killing: Landmines in Southern Africa* (New York: Human Rights Watch Arms Project, 1997).

Vollers, M., 'The Rhino Wars: Zimbabwe is Shooting Poachers Who Menace the Rare Black Rhino', *Sports Illustrated*, 2 March 1987.

Von Lettow-Vorbeck, P., *My Reminiscences of East Africa* (London: Hurst and Blackett, 1920).

Vorlaufer, K., 'CAMPFIRE – The Political Ecology of Poverty Alleviation, Wildlife Utilization and Biodiversity Conservation in Zimbabwe', *Erdkunde*, 56:2 (April–June 2002), pp. 184–206.

Wa Kinyatti, M., *Mau Mau: A Revolution Betrayed* (Jamaica, NY : Mau Mau Research Centre, 2000).

Wallace, M., *A History of Namibia: From the Beginning to 1990* (New York: Columbia University Press, 2011).

Webster, W., *Englishness and Empire, 1939–1965* (Oxford: Oxford University Press, 2005).

Wheeler, S., *Too Close to the Sun: The Audacious Life and Times of Denys Finch Hatton* (New York: Random House, 2009).

White, L., 'Poisoned Food, Poisoned Uniforms, and Anthrax: Or How Guerrillas Die in War', *Osiris*, 19 (2004), pp. 220–33.

Whyte, B., *A Pride of Eagles: the Story of Rhodesia's Air Force* (Salisbury: Graham Publishing, 1976).

Widlok, T., 'Orientation in the Wild: The Shared Cognition of HaiIom Bushpeople', *The Journal of the Royal Anthropological Institute*, 3:2 (June 1997), pp. 317–32.

Wienholt, A., *The Story of a Lion Hunt with Some of the Hunter's Military Adventures During the War* (London: Andrew Melrose, 1922).

Williams, D., *On the Border: The White South African Military Experience, 1965–1990* (Cape Town: Tafelberg, 2008).

Williams, R., 'The Other Armies: A Brief Historical Overview of Umkhonto we Sizwe (MK) 1961–1994', *South African Military History Journal*, 11:5 (June 2000).

Wilmot, A., *History of the Zulu War* (London: Richard and Best, 1880).

Wilmsen, E., *Land Filled With Flies: A Political Economy of the Kalahari* (Chicago, IL: University of Chicago Press, 1989).

Wilson, E., *Reminiscences of a Frontier Armed and Mounted Police Officer in South Africa* (Grahamstown: Campbell and Company, 1866).

Winegard, T., *Indigenous Peoples of the British Dominions and the First World War* (Cambridge: Cambridge University Press, 2012).

Wood, J. R. T. (ed.), *The War Diaries of Andre Dennison* (Gibraltar: Ashanti Publishing, 1989).

—, *A Matter of Weeks Rather Than Months: The Impasse Between Harold Wilson and Ian Smith, Sanctions, Aborted Settlements and War, 1965–69* (Victoria, BC: Trafford, 2012).

Yennaris, C., *From the East; Conflict and Partition in Cyprus* (London: Elliot and Thomson, 2003).

Yorke, E., *Mafeking, 1899–1900: Battle Story* (Stroud, UK: The History Press, 2014)

Young, J. and T. Morgan, *Animal Tracking Basics* (Mechanicsburg, PA: Stackpole Books, 2007).

NOTES

Introduction

1. L. Liebenberg, 'Persistence Hunting by Modern Hunter-Gatherers', *Current Anthropology*, Vol. 47, No. 6 (December 2006), pp. 1017–26; L. Liebenberg, *The Art of Tracking: The Origin of Science* (Claremont, South Africa: David Phillip, 1990); P. Carruthers, 'The Roots of Scientific Reasoning: Modularity and the Art of Tracking', in P. Carruthers, S. Stich, and M. Siegal (eds.), *The Cognitive Basis of Science* (Cambridge University Press, 2002), pp. 73–96; for tracking and brain size see G. Krantz, 'Brain Size and Hunting Ability in Earliest Man', *Current Anthropology*, Vol. 9, No. 5 (December 1968), pp. 450–451 and C. Brown and S. L. Lahren, 'More on Hunting Ability and Increased Brain Size', *Current Anthropology*, Vol. 14, No. 3 (June 1973), pp. 309–310; J. Sarnecki, 'The Emergence of Empathy in Cross-species Mind Reading', in L. Swan (ed.) *Origins of Mind* (London: Springer, 2013), p. 134; R. Leakey, 'Early Humans: Of Whom Do We Speak?' in F. E. Grine; J. G. Fleagle and R. E. Leakey (eds.), *The First Humans: Origin and Evolution of the Genus Homo* (London: Springer, 2009), p. 86.
2. T. R. Pickering, *Rough and Tumble: Aggression, Hunting and Human Evolution* (Berkeley: University of California Press, 2013), p. 100.
3. M. L. Hunter, R. K. Hitchcock and B. Wyckoff-Baird, 'Women and Wildlife in Southern Africa', *Conservation Biology,* Vol. 4 No. 4 (December 1990), pp. 448–451.
4. D. S. Donelan, *Tactical Tracking Operations: The Essential Guide for Military and Police Trackers* (Boulder, CO: Paladin Press, 1998); B. Carss, *The SAS Guide to Tracking* (UK: Lyons Press, 1999); L. Liebenberg, *A Photographic Guide to Tracks and Tracking in Southern Africa* (Cape Town: Struik, 2000); D. Diaz with V. L. McCann, *Tracking: Signs of Man, Signs of Hope: A Systematic Approach to the Art and Science of Tracking Humans* (Guilford, CT: Lyons Press, 2005); J. Young and T. Morgan, *Animal Tracking Basics* (Mechanicsburg, PA: Stackpole Books, 2007); L. Liebenberg, A. Louw and M. Elbroch: *Practical Tracking: A Guide to Following Footprints and Finding Animals* (Mechanicsburg, PA: Stackpole Books, 2010); J. Hurth, *Combat Tracking Guide* ((Mechanicsburg, PA: Stackpole Books, 2012); J. D. Hurth and J. W. Brokaw, 'Visual Tracking and the Military Tracking Team Capability: A Disappearing Skill and Misunderstood Capability', *Small Wars Journal*, 2010.
5. T. Brown, *The Tracker* (New York: Berkeley Books, 1978); T. Brown, *The Search: More of the Ancient Art of the New Survival* (New York; Berkeley Books, 1980); T. Brown, *The Science and Art of Tracking* (New York: Berkeley Books, 1999).
6. There is a massive literature on the history and theory of insurgency and counter-insur-

gency. J. J. McCuen, *The Art of Counter-Revolutionary War: The Strategy of Counter-Insurgency* (Harrisberg: Stackpole Books, 1966); R. G. K. Thompson, *Defeating Communist Insurgency: Experiences from Malaya and Vietnam* (Chatto and Windus, 1966); I. Beckett, *Modern Insurgencies and Counter-Insurgencies; Guerrillas and Their Opponents since 1750* (London: Routledge, 2001); J. A. Nagl, *Learning to Eat Soup with a Knife: Counter-Insurgency Lessons From Malaya and Vietnam* (University of Chicago Press, 2002); A. Mumford, *The Counter-Insurgency Myth: The British Experience of Irregular Warfare* (New York: Routledge, 2012); D. Porch, *Counter-Insurgency: Exposing the Myths of the New Way of War* (Cambridge University Press, 2013).

7.   F. Kitson, *Low Intensity Operations: Subversion, Insurgency and Peacekeeping* (London: Faber and Faber, 1971), p. 193.

8.   J. Cross, *Jungle Warfare; Experiences and Encounters* (Barnsley, UK: Pen and Sword, 2008), pp. 137, 232; I. McGibbon, *New Zealand's Vietnam War: A History of Combat, Commitment and Controversy* (Auckland: Exile Publishing, 2010), p. 178; 'British War Dogs for Vietnam', *Daily Telegraph*, 1 January 1967, p. 2; for American trackers in Vietnam and the quote see S. Merritt, *Seek On* (Strategic Book Publishing, 2011), p. 11.

9.   R. Tonsetic, *Warriors: An Infantryman's Memoir of Vietnam* (Presidio Press, 2004), p. 3.

10. Hurth and Brokaw, 'Visual Tracking'.

11. N. L. Peluso and P. Vandergeest, 'Political Ecologies of War and Forests: Counter-insurgencies and the Making of National Natures', *Annals of the Association of American Geographers*, 101 (3), 2011, pp. 587–608.

12. For more on Portuguese counter-insurgency in Africa see W. S. V. D. Waals, *Portugal's War in Angola, 1961–1974* (Pretoria: Protea Book House, 2011); J. P. Cann, *Counter-insurgency in Africa: The Portuguese Way of War, 1961–1974* (Solihull, UK: Helion and Company, 2012); A. J. Venter, *Portugal's Guerrilla Wars in Africa: Lisbon's Three Wars in Angola, Mozambique and Portuguese Guinea 1961–74* (Solihull, UK: Helion and Company, 2013).

13. L.White, 'Poisoned Food, Poisoned Uniforms, and Anthrax: Or How Guerrillas Die in War', *Osiris*, Vol. 19 (2004), pp. 220–233; Gary Baines, *South Africa's Border War: Contested Narratives and Conflicting Memories* (Bloomsbury, 2014).

14. Sisingi Kamongo and Leon Bezuidenhout, *Shadows in the Sand: A Koevoet Tracker's Story of an Insurgency War* (Johannesburg: 30 Degrees South Publishers, 2011).

1 Tracking and Identity

1.   For some works on ethnic identity in Africa see L. Vail (ed.), *The Creation of Tribalism in Southern Africa* (Berkeley, CA: University of California Press, 1991); and M. Mamdani, *Citizen and Subject: Contemporary Africa and the Legacy of Late Colonialism* (Princeton, NJ: Princeton University Press, 1996).

2.   W. Beinart and P. Coates, *Environment and History: The Taming of Nature in the USA and South Africa* (London: Routledge, 1995), p. 21.

3.   D. M. Hughes, 'Whites Lost and Found: Immigration and Imagination in Savanna Africa', in B. Caminero-Santangelo and G. Myers (eds), *Environment at the Margins: Literary and Environmental Studies in Africa* (Athens, OH: Ohio University Press, 2011), p. 165.

4.   J. Taylor, *Pondoro: The Last of the Ivory Hunters* (New York: Simon and Schuster, 1955), p. 292.

5.   J. M. MacKenzie, *The Empire of Nature: Hunting, Conservation and British Imperialism*

(New York: St Martin's Press, 1988), p. 60.

6. W. Beinart and L. Hughes, *Environment and Empire* (Oxford: Oxford University Press, 2007), pp. 62–3.

7. R. Gordon-Cumming, *Five Years of a Hunter's Life in the Far Interior of South Africa*, 2 vols (London: John Murray, 1850), vol. 1, p. 157.

8. Gordon-Cumming, *Hunter's Life*, vol. 2, p. 370.

9. Gordon-Cumming, *Hunter's Life*, vol. 1, pp. 305–6.

10. Gordon-Cumming, *Hunter's Life*, vol. 1, p. 338.

11. Gordon-Cumming, *Hunter's Life*, vol. 2, p. 29.

12. Gordon-Cumming, *Hunter's Life*, vol. 2, p. 188.

13. W. C. Baldwin, *African Hunting From Natal to the Zambesi* (London: Richard Bentley, 1863), pp. 202–3.

14. Baldwin, *African Hunting*, pp. 315–16.

15. Baldwin, *African Hunting*, p. 326.

16. Baldwin, *African Hunting*, p. 327.

17. E. Holub, *Seven Years in South Africa: Travels, Researches and Hunting Adventures Between the Diamond Fields and the Zambezi, 1872–79*, 2 vols (London: Sampson Low, 1881), vol. 1, p. 348.

18. P. Gillmore, *The Great Thirst Land: A Ride Through Natal, Orange Free State and the Kalahari Desert* (London: Cassell, Peter and Galpin, 1878), p. 291.

19. Gillmore, *Great Thirst Land*, p. 340.

20. H. Anderson Bryden, *Gun and Camera in Southern Africa* (London: Edward Stanford, 1893), p. 292.

21. F. C. Selous, *Travel and Adventure in South Eastern Africa* (London: Rowland Ward, 1893), p. 109.

22. F. C. Selous, *Travel and Adventure in South Eastern Africa* (London: Rowland Ward, 1893), p. 110.

23. S. Taylor, *The Mighty Nimrod: A Life of Frederick Courteney Selous, African Hunter and Adventurer, 1851–1917* (London: Harper Collins, 1989), p. 114.

24. For the Shikar Club see J. A. Mangan and C. McKenzie, *Militarism, Hunting, Imperialism: 'Blooding' the Martial Male* (New York: Routledge, 2010), pp. 168–92; and M. Rico, *Nature's Noblemen: Trans-Atlantic Masculinities and the Nineteenth Century American West* (New Haven, CT: Yale University Press, 2013), pp. 196–7.

25. F. R. Burnham, *Scouting on Two Continents* (Garden City, NY: Garden City Publishing, 1926), p. 320.

26. A. Wienholt, *The Story of a Lion Hunt with Some of the Hunter's Military Adventures During the War* (London: Andrew Melrose, 1922), p. 40.

27. S. S. Dornan, 'The Tati Bushmen (Masarwas) and their Language', *The Journal of the Royal Anthropological Institute of Great Britain and Ireland*, 47 (January–June 1917), pp. 37–112, on pp. 39, 45.

28. J. H. Venning, 'A Lesson in Lion Hunting', *Northern Rhodesia Journal*, 3:5 (1958), pp. 425–7.

29. J. D. Clark, 'Bushmen Hunters of the Barotse Forests', *Northern Rhodesia Journal*, 1:3 (1951), pp. 56–65, on p. 61.

30. L. Marshall, 'Marriage Among !Kung Bushmen', *Africa: Journal of the International African Institute*, 29:4 (October 1959), pp. 335–65, on p. 360.

31. E. M. Thomas, *The Old Way: A Story of the First People* (New York: Picador, 2006), p. 99.

32. P. V. Tobias and M. Biesele, *The Bushmen: San Hunters and Herders of Southern Africa* (Cape Town: Human and Rousseau, 1978), p. 26.

33. R. B. Lee, *The! Kung San: Men, Women, and Work in a Foraging Society* (Cambridge: Cambridge University Press, 1979), p. 212.

34. L. V. Vuuren, 'The Many Myths of Laurens van der Post: Van Der Post and Bushmen in the Television Series Lost World of Kalahari (1958)', *South African Historical Journal*, 48:1 (2003), pp. 47–60; and L. V. Vuuren, '"And He Said They Were Ju/Wasi, the People...": History and Myth in John Marshall's "Bushman Films" 1957–2000', *Journal of Southern African Studies*, 35:3 (2009), pp. 557–74.

35. T. Widlok, 'Orientation in the Wild: The Shared Cognition of HaiIIom Bushpeople', *The Journal of the Royal Anthropological Institute*, 3:2 (June 1997), pp. 317–32.

36. H. Reuning and W. Wortley, 'Psychological Studies of the Bushman', *Psychologica Africana*, 7 (1973), pp. 1–113.

37. 'Tracking with the Bushmen of the Kalahari', *Natural High Safaris*, at http://www.naturalhighsafaris.com/browse/experience/tracking_with_the_bushmen_of_the_kalahari [accessed 13 April 2015].

38. G. B. Silberbauer, 'Note on Proposed Game Reserve and Bushmen', 19 May 1960, Botswana National Archives (hereafter BNA), Game Reserve, Central Kalahari, Ghanzi District, S.580/7/1, 1963.

39. G. B. Silberbauer, 'Masarwa Affairs and Bushman Survey', 24 February 1960, Ghanzi to Government Secretary, Mafeking, BNA, S563/1/2.

40. P. H. Capstick, *Death in the Long Grass: A Big Game Hunter's Adventures in the African Bush* (New York: St. Martin's Press, 1977), p. 116.

41. Capstick, *Death in the Long Grass*, p. 118.

42. M. Main, 'Notes on Bushman Tracking Techniques', unpublished paper, 22 May 2009. I would like to thank Mike Main for providing me with a copy of his paper.

43. W. R. Foran, *Kill or be Killed: The Rambling Reminiscences of an Amateur Hunter* (London: Hutchinson, 1933), p. 122.

44. Foran, *Kill or Be Killed*, p. 123.

45. T. Roosevelt, *African Game Trails: An Account of the African Wanderings of an American Hunter-Naturalist* (New York: Charles Scribner and Sons, 1921), p. 362.

46. E. J. G. Huxley, *The Mottled Lizard* (London: Chatto and Windus, 1962), p. 211.

47. W. P. Lineberry, *East Africa* (New York: W.H. Wilson, 1968), p. 90.

48. S. Wheeler, *Too Close to the Sun: The Audacious Life and Times of Denys Finch Hatton* (New York: Random House, 2009), p. 53.

49. W. S. Rainsford, *The Land of the Lion* (New York: Doubleday, Page and Coy, 1909), p. 110.

50. C. H. Stigand, *The Game of British East Africa* (London: Harper Cox, 1909), p. 136; see also E. Steinhart, *Black Poachers, White Hunters: A Social History of Hunting in Colonial Kenya* (Oxford: James Currey, 2006), p. 76.

51. A. Blayney Percival, *A Game Ranger's Notebook* (London: Nisbet, 1924), p. 324.

52. J. Stevenson-Hamilton, *The Low-Veld: Its Wild Life and Its People* (London: Cassell and Company, 1927), p. 215.

53. K. Meadows, *Sometimes When It Rains: White Africans in Black Africa* (Bulawayo, Zimbabwe: Thorntree Press, 2000), p. 282.

54. F. van Riel, *My Life with Leopards: Graham Cooke's Story* (South Africa: Penguin Books, 2013), electronic copy.

55. R. Frump, *The Man-Eaters of Eden: Life and Death in Kruger National Park* (Guilford, CT: The Lyons Press, 2006), p. 61.

56. S. Robins and K. V. D. Waal, 'Model Tribes and Iconic Conservationists? Tracking the Makuleke Restitution Case in Kruger National Park', in C. Walker, A. Bohlin, R. Hall and T. Kepe (eds), *Land, Memory, Reconstruction, and Justice: Perspectives on Land Claims in South Africa* (Athens, Ohio: Ohio University Press, 2010), pp. 163–80, on p. 166.

57. 'Simbavati River Lodge', *Bushscapes*, at http://bushscapes.co.za/greater-kruger-park/timbavati/simbavati-river-lodge/ [accessed 13 April 2015].

58. J. Sevor, 'Shangaan Bull Zimbabwe', *Africa Hunting*, 13 January 2010, at http://www.africahunting.com/content/2-shangaan-bull-zimbabwe-437/ [accessed 13 April 2015].

59. S. Foran, 'Graduate Student Uses Ancient Skills in Modern Conservation Efforts', *UCONN Today*, 12 June 2013, at http://today.uconn.edu/blog/2013/06/graduate-student-uses-ancient-skills-in-modern-conservation-efforts/ [accessed 13 April 2015].

60. P. H. G. Powell-Cotton, 'Notes on a Journal Through the Great Ituri Forest', *Journal of the Royal African Society*, 7:25 (October 1907), pp. 1–12, on p. 5; see also P. H. G. Powell-Cotton, 'A Journey Through the Eastern Portion of the Congo State', *The Geographic Journal*, 30:4 (October 1907), pp. 371–82.

61. P. J. Pretorius, *Jungle Man* (1947; Alexander, NC: Alexander Books, 2001), p. 104.

62. C. Christy, 'The Ituri River, Forest and Pygmies', *The Geographic Journal*, 46:3 (September 1915), pp. 205–6.

63. B. Burbridge, *Tracking and Capturing the Ape-Man of Africa* (New York: The Century Company, 1928), p. 212.

64. A. Gatti, *Great Mother Forest* (New York: Charles Scribner's Sons, 1937), p. 104.

65. S. Jones, *Under the African Sun* (London: Hurst and Brackett, 1956), p. 191.

66. P. Flack, 'Rainforest Elephant Hunting in Cameroon', at http://www.peterflack.co.za/articles/articleRainForestElephantHunt.html [accessed 13 April 2015].

67. C. Hopkins, 'Tarzan Had it Wrong', at www.http://cameronhopkinsblog.blogspot.ca/2006/08/tarzan-had-it-all-wrong.html [accessed 13 April 2015].

68. 'Gorilla Tracking', *Dzanga Sangha National Park*, at http://www.dzanga-sangha.org/node/326 [accessed 13 April 2015].

69. C. Cipolletta, 'The Long Road to Habituation: A Window into the Lives of Gorillas', in M. M. Robbins and C. Boesch (eds), *Among African Apes: Stories from the Field* (Berkeley, CA: University of California Press, 2011), pp. 129–42, on p. 131.

70. J. Lewis, *The Batwa Pygmies of the Great Lakes Region*, Report of Minority Rights Group International (London: MRG, 2000), p. 25.

71. P. Gillmore, *The Hunter's Arcadia* (London: Chapman and Hall, 1886), p. 187.

72. T. J. Stapleton, Interview with A. Campbell, Gaborone, Botswana, 10 August 2010.

73. F. J. Bagshawe, 'The Peoples of the Happy Valley (East Africa) The Aboriginal Races of Kondoa Irangi Part II: The Kangeju', *Journal of the Royal African Society*, 24:94 (January 1925), pp. 117–30, on p. 123.

74. C. M. Doke, *The Lambas of Northern Rhodesia: A Study of Their Customs and Beliefs* (London: George G. Harrap, 1931), p. 334.

75. P. V. Lettow-Vorbeck, *My Reminiscences of East Africa* (London: Hurst and Blackett, 1920), p. 10.

76. Burbridge, *Gorilla*, pp. 72, 176.

77. J. A. Hunter, *Hunter's Tracks* (London: Hamish Hamilton, 1957), pp. 126–7.

78. G. H. Anderson, *African Safaris* (Edmonton: Safari Press, 2000), p. 15.

79. C. J. P. Ionides, *Mambas and Man-Eaters: A Hunter's Story* (New York: Holt, Rinehart and Winston, 1966), p. 168.

80. N. Steele, *Bushlife of a Game Warden* (Cape Town: T. V. Bulpin, 1979), p. 210.

81. I. Player, *Zulu Wilderness: Shadow and Soul* (Golden, CO: Fulcrum Publishing, 1998), p. 122.
82. R. Mears, *My Outdoor Life* (London: Hodder and Stoughton, 2013), p. 332.

2 Tracking and Colonial Warfare

1. R. K. Hitchcock, 'Resource Depletion and Coercive Conservation Among the Tyua of the Northeastern Kalahari', *Human Ecology*, 23:2 (June 1995), p. 179; and T. Stapleton, *Faku: Rulership and Colonialism in the Mpondo Kingdom, c. 1780–1867* (Waterloo, Ontario: Wilfrid Laurier University Press, 2001), p. 127.
2. I. Player, *Zulu Wilderness: Shadow and Soul* (Golden, CO: Fulcrum Publishing, 1998), p. 13.
3. T. W. Dunlay, *Wolves for the Blue Soldiers: Indian Scouts and Auxiliaries with the United States Army* (Lincoln, NE: University of Nebraska Press, 1982), p. 78.
4. Dunlay, *Wolves for the Blue Soldiers*, p. 90.
5. T. Goodrich, *Black Flag: Guerilla Warfare on the Western Border, 1861–65* (Bloomington, IN: Indiana University Press, 1999), p. 48; and J. C. Inscoe and G. B. McKinney, *The Heart of Confederate Appalachia: Western North Carolina in the Civil War* (Chapel Hill, NC: University of North Carolina Press, 2000), p. 123.
6. J. Richards, *The Secret War: A True History of Queensland's Native Police* (St Lucia: University of Queensland Press, 2008); T. Winegard, *Indigenous Peoples of the British Dominions and the First World War* (Cambridge: Cambridge University Press, 2012), p. 48; and A. L. Haydon, *The Trooper Police of Australia: A Record of Mounted Police Work in the Commonwealth from the Earliest Days of Settlement to the Present Time* (London: Andrew Melrose, 1911), p. 197.
7. N. Mostert, *Frontiers: The Epic of South Africa's Creation and the Tragedy of the Xhosa People* (New York: Alfred Knopf, 1992), pp. 450, 627–8, 707.
8. C. A. Fitzroy, Captain Horse Guards, Military Secretary, 'Instructions for the Commandant on the Frontier', 25 March 1825, *Papers Relative to the Condition and Treatment of the Native Inhabitants of Southern Africa, Within the Colony of the Cape of Good Hope, or Beyond the Frontier of that Colony*, Part I: Hottentots and Bosjesmen; Caffres; Griquas, House of Commons, British Parliamentary Paper (hereafter BPP) 50, 1835, p. 178.
9. Colonel H. Somerset to Commissioners of Inquiry, 6 February 1826, in *Papers Relative to the Condition and Treatment of the Native Inhabitants of Southern Africa*, Part II, BPP 252, 1835, p. 148.
10. Evidence of Colonel T. F. Wade, 25 April 1836, *Report from the Select Committee on Aborigines (British Settlements); together with the Minutes of Evidence, Appendix and Index*, BPP 583, 1836, p. 429.
11. Field Cornet L. Nel to Lieutenant Colonel Somerset, 30 June 1834, *Caffre War*, BPP 503, 1837, p. 137.
12. Thomas Phillips to Lord Stanley, Secretary of State for Colonies, 28 July 1842, *Correspondence with the Governor of the Cape of Good Hope Relative to the State of the Kafir Tribes*, BPP 424, 1851, p. 128.
13. Lieutenant Colonel T. Wade, Acting Governor to Secretary Stanley, 14 January 1834, *Papers Relative to the Condition and Treatment of the Native Inhabitants of Southern Africa*, BPP 1835 (252), part II, p. 78.
14. J. E. Alexander, *Narrative of a Voyage of Observation Among the Colonies of Western Africa, in the Flag-ship Thalia, and a Campaign in Kaffir-Land* (London: Henry Colburn, 1837), p. 28.

15. Alexander, *Narrative of a Voyage of Observation*, p. 36.
16. Evidence of Rev. William Shaw, 7 August 1835, *Report from the Select Committee on Aborigines (British Settlements)*, BPP 1836 (538), p. 63.
17. Alexander, *Narrative*, p. 49.
18. Colonel George MacKinnon to Lieutenant General Sir H. B. Smith, High Commissioner, 7 January 1850, *Correspondence with the Governor of the Cape of Good Hope Relative to the State of the Kafir Tribes on the Eastern Frontier of the Colony*, BPP 1850 (1288), p. 25.
19. H. B. Smith to Earl Grey, 17 March 1851, *Correspondence with the Governor of the Cape of Good Hope Relative to the State of the Kafir Tribes*, BPP 1851 (424), p. 22.
20. Statement of Major J. E. Alexander, 3 July 1851, Report from the Select Committee on the Kafir Tribes, BPP, 2 August 1851, p. 345.
21. Sir H. H. Parr, *A Sketch of the Kafir and Zulu Wars: From Guadana to Isandhlwana* (London: C. Kegan Paul, 1880), p. 29.
22. E. Wilson, *Reminiscences of a Frontier Armed and Mounted Police Officer in South Africa* (Grahamstown: Campbell and Company, 1866), p. 52.
23. Wilson, *Reminiscences*, p. 139.
24. Alexander, *Narrative*, p. 25.
25. J. Milton, *The Edges of War: A History of Frontier Wars (1702–1878)* (Cape Town: Juta, 1983), p. 164.
26. D. C. F. Moodie, *The History of the Battles and Adventures of the British, the Boers and the Zulus in Southern Africa* (Sydney: George Robertson, 1879), p. 93.
27. A. Wilmot, *History of the Zulu War* (London: Richard and Best, 1880), p. 130.
28. Colonel Schermbrucker to Colonel E. Wood, 11 February 1879, *Further Correspondence Respecting the Affairs of South Africa*, BPP, March 1879, p. 65.
29. P. Thompson, *Black Soldiers of the Queen: The Natal Native Contingent in the Anglo-Zulu War* (Tuscaloosa, AL: University of Alabama Press, 2006), p. 161.
30. F. R. Burnham, *Scouting on Two Continents* (Garden City, NY: Garden City Publishing, 1926), p. 163.
31. F. R. Burnham, *Scouting on Two Continents* (Garden City, NY: Garden City Publishing, 1926), pp. 191–2, on p. 197.
32. B. A. le Cordeur (ed.), *The Journal of Charles Lennox Stretch* (Grahamstown: Longman, 1988), p. 109.
33. Alexander, *Narrative*, p. 37.
34. Alexander, *Narrative*, pp. 350–1, extract of a letter from Lieutenant Moultie to Sir. B. D'Urban, 12 September 1835.
35. Lieutenant Colonel T. Fordyce to H. L. Maydwell, Military Secretary, 17 October 1851, *Correspondence with the Governor of the Cape of Good Hope Relative to the State of the Kafir Tribes*, BPP 1852 (1428), p. 183.
36. A. J. Cloete, 'Summary of Military Events Connected with the Operations of the Army in the Field, from the 12th October to the 12th November', Grahamstown, 12 November 1852, *Correspondence with the Governor of the Cape of Good Hope Relative to the State of the Kafir Tribes*, BPP 1852–53 (1635), p. 198.
37. C. Hummel (ed.), *The Frontier War Journal of Major John Crealock 1878* (Cape Town: Van Riebeeck Society, 1988), p. 53.
38. I. Castle, *The British Infantryman in South Africa, 1877–81* (Oxford: Osprey, 2003), p. 42.
39. F. C. Selous, *Sunshine and Storm in Rhodesia* (London: Rowland Ward, 1896), p. 121.
40. R. S. S. Baden-Powell, *The Matabele Campaign 1896* (London: Methuen, 1900), p. 32.

41. R. S. S. Baden-Powell, *Cavalry Instruction* (London: Harrison and Sons, 1885), p. 86.
42. Baden-Powell, *Matabele*, pp. 33, 54.
43. Baden-Powell, *Matabele*, p. 40.
44. R. MacDonald, *Sons of Empire: The Frontier and the Boy Scout Movement, 1890–1918* (Toronto: University of Toronto Press, 1993), p. 70.
45. T. Jeal, *Baden-Powell: Founder of the Boy Scouts* (New Haven, CT: Yale University Press, 2007), p. 188.
46. D. Reitz, *Command: A Boer Journal of the Boer War* (London: Faber and Faber, 1929). p. 223.
47. W. Storey, *Guns, Race and Power in Colonial South Africa* (Cambridge: Cambridge University Press, 2008), p. 321.
48. Burnham, *Scouting*, p. 142.
49. E. Yorke, *Mafeking, 1899–1900: Battle Story* (Stroud: The History Press, 2014), pp. 91–2.
50. B. Nasson, *Abraham Esau's War: A Black South African's War in the Cape, 1899–1902* (Cambridge: Cambridge University Press, 1991), pp. 23, 97.
51. P. Robinson, 'The Search for Mobility During the Second Boer War', *Journal of the Society for Army Historical Research*, 68:346 (2008), pp. 140–57, on p. 149.
52. H. Quan, *Sam Steele: The Wild West Adventures of Canada's Most Famous Mountie* (Victoria, BC: Heritage House Publishing, 2003); R. Stewart, *Sam Steele: Lion of the Frontier* (Toronto: Doubleday, 1979); and C. Miller, *Painting the Map Red: Canada and the South African War* (Toronto: McGill-Queen's Press, 1993).
53. D. Huggonson, 'The Black Trackers of Bloemfontein', *Land Rights News*, February 1990, p. 20; Winegard, *Indigenous Peoples*, p. 57; D. Kerwin, 'The Lost Trackers: Aboriginal Servicemen in the 2nd Boer War', *Sabretache*, 54:1 (March 2013), pp. 4–14; and S. Casey, 'Tracking the Trackers: Hunt Begins for Abandoned Bushmen', *Brisbane Times*, 7 September 2009.
54. C. E. Callwell, *Small Wars: Their Principles and Practise* (1896; Lincoln, NE: University of Nebraska Press, 1996), p. 144.
55. R. S. S. Baden-Powell, *Aids to Scouting for N.C.O.'s and Men* (1899; Aldershot: Gale and Polden, 1915); and R. S. S. Baden-Powell, *Scouting for Boys* (London: Pearson, 1908).
56. T. Parsons, *Race, Resistance and the Boy Scout Movement in British Colonial Africa* (Athens, OH: Ohio University Press, 2004), p. 78.
57. MacDonald, *Sons of Empire*, pp. 54–9; G. A. Pocock, *Outrider of Empire: The Life and Adventures of Roger Pocock* (Edmonton, Alberta: University of Alberta Press, 2007).
58. S. Jones, *From Boer War to World War: Tactical Reform of the British Army, 1902–1914* (Norman, OK: University of Oklahoma Press, 2012), p. 201.
59. S. Wheeler, *Too Close to the Sun: The Audacious Life and Times of Denys Finch Hatton* (New York: Random House, 2009), p. 82.
60. N. G. Speed, *Born to Fight: Major Charles Joseph Ross, A Definitive Study of His Life* (Melbourne: The Caps and Flints Press, 2002).
61. A. Blayney-Percival, *A Game Ranger on Safari* (London: Nisbet, 1928), p. 287.
62. R. Anderson, *The Forgotten Front: The East African Campaign, 1914–1918* (Stroud: Tempus, 2004), pp. 91, 257; and P. H. Capstick, *Warrior: The Legend of Colonel Richard Meinertzhagen* (New York: St Martin's Press, 1998), pp. 197–200.
63. Pretorius, *Jungle Man*, p. 165.
64. M. T. Hoffman, 'Major P. J. Pretorius and the Decimation of the Addo Elephant Herd in 1919–20: Important Reassessments', *Koedoe*, 36:2 (1993), pp. 23–44.
65. Capstick, *Warrior*, p. 184.

66. E. Mandiringana and T. Stapleton, 'The Literary Legacy of Frederick Courteney Se-lous', *History in Africa*, 25 (1998), pp. 199–218.

67. E. Paice, *Tip and Run: The Untold Tragedy of the Great War in Africa* (London: Wei-denfeld and Nicolson, 2007), pp. 179, 195; and B. M. Du Toit, *The Boers in East Africa: Ethnicity and Identity* (Westport, CT: Greenwood, 1998), pp. 96, 180.

68. F. J. Nusselhuf, 'General Paul Von Lettow-Vorbeck's East Africa Campaign: Maneuver Warfare on the Serengeti' (MA Thesis, University of North Texas, 2012), pp. 16, 26.

69. A. Wienholt, *The Story of a Lion Hunt with Some of the Hunter's Military Adventures During the War* (London: Andrew Melrose, 1922); and P. V. Lettow-Vorbeck, *My Reminiscences of East Africa* (London: Hurst and Blackett, 1920), p. 143.

70. E. Wilmsen, *Land Filled With Flies: A Political Economy of the Kalahari* (Chicago, IL: University of Chicago Press, 1989), p. 148.

71. R. Gordon and S. S. Douglas, *The Bushman Myth: The Making of a Namibian Under-class* (Boulder, CO: Westview Press, 2000), p. 130; R. J. Gordon, '"Captured on Film": Bushmen and the Claptrap of Performative Primitives', in P. Landou and D. Kaspin (eds), *Images and Empire: Visuality in Colonial and Post-Colonial Africa* (Berkeley, CA: University of California Press, 2002), p. 224.

72. U. Dieckmann, *Haillom in the Etosha Region: A History of Colonial Settlement, Ethnic-ity and Nature Conservation* (Basel, Switzerland: Basler Afrika, 2007), p. 183.

3 Kenya, 1952–6

1. I. Parker, *The Last Colonial Regiment: The History of the Kenya Regiment (TF)* (Moray: Librario Publishing, 2009), pp. 27–48.

2. For the colonial view see L. S. B. Leakey, *Mau Mau and the Kikuyu* (London: Rout-ledge, 1952); and F. D. Corfield, *The Origins and Growth of Mau Mau: An Historical Survey* (Nairobi: Colony and Protectorate of Kenya, 1960). For the national view see C. G. Rosberg and J. C. Nottingham, *The Myth of 'Mau Mau': Nationalism in Kenya* (New York: Praeger, 1966); and M. W. Kinyatti, *Mau Mau: A Revolution Betrayed* (New York: Mau Mau Research Centre, 2000). For the academic view see R. Buijten-huijs, *Mau Mau: Twenty Years After: The Myth and the Survivors* (New York: Mouton, 1973); D. Throup, *Economic and Social Origins of Mau Mau, 1945–53* (Oxford: James Currey, 1987); F. Furedi, *The Mau Mau War in Perspective* (Oxford: James Currey, 1989); and B. Berman, and J. Lonsdale, *Unhappy Valley: Conflict in Kenya and Africa* (London: James Currey, 1992).

3. For recent studies of the conflict see D. Branch, *Defeating Mau Mau, Creating Kenya: Counterinsurgency, Civil War and Decolonization* (Cambridge: Cambridge University Press, 2009); H. Bennett, *Fighting the Mau Mau: The British Army and Counter-Insur-gency in the Kenya Emergency* (Cambridge: Cambridge University Press, 2013); and S. Chappell, 'Air Power in the Mau Mau Conflict: The Government's Chief Weapon', *The RUSI Journal*, 156:1 (February-March 2011), pp. 64–70.

4. D. Anderson, *Histories of the Hanged: The Dirty War in Kenya and the End of Empire* (London: Weidenfeld and Nicolson, 2005); C. Elkins, *Imperial Reckoning: The Untold Story of Britain's Gulag in Kenya* (New York: Henry Holt, 2005); and J. Blacker, 'The Demography of Mau Mau: Fertility and Mortality in Kenya in the 1950s: A Demogra-pher's Viewpoint', *African Affairs*, 106 (2007), pp. 205–27.

5. E. Steinhart, *Black Poachers, White Hunters: A Social History of Hunting in Colonial Kenya* (Oxford: James Currey, 2006).

6. C. L. Hiscox, *The Dawn Stand-to: The Life of IVB (Peter) Mills, QPM, CPM* (Bideford: Edward Gaskell Publishers, 2000), p. 280.

7. Hiscox, *Dawn Stand-to*, p. 278.

8. Parker, *Last Colonial Regiment*, p. 241.

9. Major General Heyman, Chief of Staff, 'General Headquarters East Africa, Training Instruction No. 9, Patrols and Ambushes', 2 December 1953, National Archives, UK, WO 276/249.

10. Parker, *Last Colonial Regiment*, pp. 207, 343–4.

11. A. House, *The Great Safari: The Lives of George and Joy Adamson* (New York: William Morrow, 1993), pp. 197–209.

12. House, *The Great Safari*, p. 206.

13. F. Bartlett, *Shoot Straight and Stay Alive: A Lifetime of Hunting Experiences* (Johannesburg: Rowland Ward Publications, 1994), p. 47.

14. D. Holman, *Elephants at Sundown: The Story of Bill Woodley* (London: W. H. Allen, 1978), pp. 71–3. For Hekuta Simba's father (called Simba) see D. Holman, *The Elephant People* (London: James Murray, 1967), p. 94.

15. W. Webster, *Englishness and Empire, 1939–1965* (Oxford: Oxford University Press, 2005), p. 125. See also S. Harrison, *Dark Trophies: Hunting and the Enemy Body in Modern War* (New York: Berghahn Books, 2012), pp. 160–1.

16. Parker, *Last Colonial Regiment*, p. 273.

17. Bartlett, *Shoot Straight*, p. 85.

18. Parker, *Last Colonial Regiment*, pp. 209–21, 267.

19. Parker, *Last Colonial Regiment*, p. 229.

20. G. Campbell, *The Charging Buffalo: A History of the Kenya Regiment, 1937–1963* (London: Leo Cooper, 1986), p. 58.

21. M. Brown, 'The Tracking School – Nanyuki (1953–54)', *Buffalo Barua: The Newsletter of the Kenya Regiment Association of Europe and North America*, November 2007, pp. 35–40.

22. Parker, *Last Colonial Regiment*, pp. 229–35. For the details of Woodley's recruiting see Holman, *Elephants at Sundown*, p. 74.

23. C. Hornsby, *Kenya: A History Since Independence* (London: I. B. Tauris, 2013), p. 46 mistakenly claims that most security force trackers were Samburu. For military recruitment of imagined martial people see T. H. Parsons, *The African Rank-and-File: Social Implications of Colonial Military Service in the King's African Rifles, 1902–1964* (Oxford: James Currey, 1999), pp. 53–103; and for Kamba in the safari industry see Steinhart, *Black Poachers, White Hunters*, p. 57.

24. Parker, *Last Colonial Regiment*, p. 234.

25. S. Jones, *Under the African Sun* (London: Hurst and Brackett, 1956), p. 191.

26. *Parker, Last Colonial Regiment*, p. 215.

27. Recorded interview with M. G. L. Potts by N. de Lee, 14 June 2002, Imperial War Museum, 23213.

28. Campbell, *Charging Buffalo*, p. 52.

29. J. A. Rutherford, 'History of the Kikuyu Guard', Imperial War Museum, London, Misc. 134, Item 2084, p. 12.

30. R. G. Turnbull, Government Secretary, 'Award of the British Empire Medal', 26 January 1956, *Kenya Gazette*, 7 February 1956, p. 88; and 'Extract from a Letter by Mervyn Swire Ray, 5 November 1956', at http://www.goosefamily.org/families/ray/notes/006. htm [accessed 15 April 2015]. For the quotation see V. Fey, *Cloud Over Kenya*

(London: Collins, 1964), p. 123. Gichimu's honey collecting skills are described in E. Huxley, *Nine Faces of Kenya* (London: Harvill Press, 1991).

31. M. Brown, 'The Tracking School – Nanyuki (1953–54)', *Buffalo Barua: The Newsletter of the Kenya Regiment Association of Europe and North America*, November 2007, pp. 35–40; Parker, *Colonial Regiment*, pp. 245–246; and 'Recommendation for Award for Tooley, James Peter', 31 January 1956, National Archives, UK, WO 373/121/40.

32. T. Moreman, *The Jungle, the Japanese and the British Commonwealth Armies at War 1941–45: Fighting Methods, Doctrine and Training for Jungle Warfare* (New York: Frank Cass, 2005), p. 93.

33. L. Gill, *Remembering the Regiment* (Victoria, British Columbia: Trafford, 2004), p. 2.

34. Brown, 'The Tracking School'; and Parker, *Colonial Regiment*, p. 246.

35. Interview with Potts.

36. Gill, *Remembering*, p. 76.

37. Gill, *Remembering*, p. 60.

38. Interview with Potts.

39. 'A Handbook on Anti-Mau Mau Operations', East Africa Command, n.d., p. 11. See also 'Training for Forest Operations', Kenya Regiment, 1953, from private collection of John Davies.

40. Chief of Staff to Member for Finance and Development, Nairobi, 'African Trackers', 13 July 1954, National Archives, UK, WO 276/248; and Butler for Commander-in-Chief to Minister for Finance and Development, Nairobi, 'African Trackers', 17 December 1954.

41. J. J. Hespeler-Boultbee, *Mrs. Queen's Chump: Idi Amin, Communists and Other Silly Follies of the British Empire: A Military Memoir* (British Columbia, Canada: CCB Publishing, 2012), p. 93.

42. C. Wood, Recorded interview with Peter Holmes Walter Brind, Imperial War Museum, 10089.

43. Lieutenant Colonel AAG to Command Secretary (Mr. White), 11 May 1955, National Archives, UK, WO 276/249; Lieutenant Colonel GSO I (Ops) to Chief of Staff, 25 June 1955, National Archives, UK, WO 276/249; and Major GSO 2 (Ops and Trg) to GSO I, 'Trackers', 13 July 1955, National Archives, UK, WO 276/249.

44. Parker, *Colonial Regiment*, pp. 289–94; Bennett, *Fighting the Mau Mau*, pp. 152–8; and F. Kitson, *Gangs and Counter-Gangs* (London: Barrie, 1960).

45. Campbell, *The Charging Buffalo*, p. 59; and Holman, *Elephants at Sundown*, p. 80. There is some confusion over the identity of the African tracker as Holman, quoting Woodley, and Campbell says it was Gibson Wambugu who was one of Woodley's Waata trackers.

46. D. Sheldrick, *An African Love Story: Love, Life and Elephants* (London: Penguin, 2012), p. 63.

47. House, *The Great Safari*, pp. 118–19; and Holman, *Elephants at Sundown*, p. 58.

48. Hiscox, *Dawn Stand-to*, p. 286.

49. R. W. Heather, 'Of Men and Plans: The Kenya Campaign as Part of the British Counter-Insurgency Experience', *Conflict Quarterly* (Winter 1993), p. 20; Bennett, *Fighting the Mau Mau*, p. 27; H. Bennett and R. Cormac, 'Low Intensity Operations in Theory and Practise: General Sir Frank Kitson as Warrior-Scholar', in A. Mumford and B. C. Reis (eds), *The Theory and Practise of Irregular Warfare: Warrior Scholarship in Counter-Insurgency* (New York: Routledge, 2014), pp. 105–24, on p. 109.

50. Major General Heyman, Chief of Staff, 'General Headquarters East Africa, Training Instruction No. 9, Patrols and Ambushes', 2 December 1953, National Archives, UK,

WO 276/249.

51. Elliott, 'Notes on Tracker Combat Teams', [*c.* July or August 1954], National Archives, UK, WO 276/248.

52. 49 Brigade commander to General Headquarters, East Africa, 'Tracking Combat Teams', 30 July 1954, National Archives, UK, WO 276/248; and Elliott to Colonel R. Butler, GSO Intelligence, 'Combat Tracker Teams', 8 August 1954, National Archives, UK, WO 276/248.

53. Major General Heyman, Chief of Staff to Brigade Commanders, 'Training of Tracking Combat Teams', 8 July 1954, National Archives, UK, WO 276/248.

54. Major General G. D. G. Heyman, General Headquarters, East Africa to Brigadier George Taylor, 49 Brigade, 16 August 1954, National Archives, UK, WO 276/248.

55. For quotes see Parker, *Colonial Regiment*, p. 266;

56. 49 Brigade Commander to General Headquarters, East Africa, 'Tracker Combat Teams: Mr. Venn Fey', 30 August 1954, National Archives, UK, WO 276/248; and Major General Heyman, Chief of Staff to Kenya Regiment, 'Commission', 31 August 1954, National Archives, UK, WO 276/248. For Gichimu see V. Fey, *Wide Horizon: Tales of a Kenya That Has Passed into History* (New York: Vantage Press, 1982), pp. 67, 70.

57. Captain V. Fey, 'Report on Patrols Carried Out By 49 Brigade Combat Tracker Group from 29 September to 8 October', 10 October 1954, National Archives, UK, WO 276/248; Lieutenant Colonel GSO1 to Chief of Staff, '49 Brigade Combat Tracker Teams', 16 October 1954, National Archives, UK, WO 276/248; Captain V. Fey, 'Report on Patrol Carried Out By 49 Brigade Combat Tracker Teams Along the Forest Reserve Edge Bordering Loc 16, Loc 2 Fort Hall, 18 to 31 October 1954', 31 October 1954, National Archives, UK, WO 276/248.

58. 'Success of Tracker Teams in Kenya', *The Times*, 1 November 1954, p. 4.

59. Heyman to Brigadier Orr, 70 Brigade, 15 November 1954; Heyman to Brigadier Taylor, 49 Brigade, 15 November 1954, National Archives, UK, WO 276/248; and Heyman to Brigadier the Lord Thurlow, 39 Brigade, 15 November 1954, National Archives, UK, WO 276/248. For Elliott on the KAR see Elliott to Colonel Butler, 1 August 1954, National Archives, UK, WO 276/248.

60. Lieutenant Colonel GSO1 Ops, 'Specialist Forces to Combat Mau Mau', 25 November 1954, National Archives, UK, WO 276/248; and Lieutenant Colonel GSO1 Ops, 'Operations After Hammer – Draft Directive for C in C', 4 December 1954, National Archives, UK, WO 276/248.

61. For Hammer see I. Beckett, *Modern Insurgencies and Counter-Insurgencies: Guerrillas and Their Opponents* (London: Routledge, 2001), p. 126. For Fey's work during Hammer see 'Recommendation for Award for Fey, Venn', 19 July 1955, National Archives, UK, WO 373/121/16; and Major Guy Campbell, Kenya Regiment to General Headquarters, East Africa, 'Forest Operating Companies', 12 January 1955, National Archives, UK, WO 276/249.

62. Major General Heyman, Chief of Staff to Brigades, 'Training of Forest Operating Companies and Trojan Teams', 30 December 1954, National Archives, UK, WO 276/249; Lieutenant Colonel GSO1 Ops to DADV&RS, 'War Dogs', 3 January 1955; Heyman to Brigades, 25 March 1955, National Archives, UK, WO 276/249.

63. V. Fey, *Cloud Over Kenya* (London: Collins, 1964), p. 17.

64. 'Kenya, Press Extracts, January 1955', University of the Witwatersrand, Historical Papers.

65. M. Page, *KAR: A History of the King's African Rifles* (London: Leo Cooper, 1998), p. 205.

66. 'Anti-Mau Mau Operations', p. 11.

67. Major A. J. Wilson, 'Forming a Forest Operating Company', The Rifle Brigade Chronicles, 1954–57, at http://www.greenjackets-net.org.uk/rb/kenmala.html [accessed 15 April 2015].
68. Parker, *Last Colonial Regiment*, p. 271.
69. 'Unorthodox Methods Against Forest Terrorists: Recruitment of Tribal Tracker Teams', *The Times*, 21 June 1955, p. 9.
70. Bennett, *Fighting the Mau Mau*, p. 155.
71. Captain R. J. Folliott, 'Special Force Patrol Report', 22 October 1955 and 23 January 1956, National Archives, UK, WO 276/431.
72. Lieutenant Colonel GSO I to Chief of Staff, 'Trackers', 25 June 1955, National Archives, UK, WO 276/249; Major GSO 2 (Ops and Trg) to GSO I (East Africa), 'Trackers', 13 July 1955, National Archives, UK, WO 276/249; 'Notes on C-in-C's Tour, 29–30 June 1955', National Archives, UK, WO 276/249. Bennett, *Fighting the Mau Mau*, p. 153 mistakenly refers to the 205 trackers as all former Mau Mau.
73. Ministry of Defence, 'Police Tracker Teams – Nanyuki District', 20 June 1955, National Archives, UK, WO 276/249; G. R. H. Gribble for Commissioner of Police to Secretary of Defence, 'Stock Thefts – Mweiga/Ngobit', 24 July 1954, National Archives, UK, WO 276/248; and Assistant Commissioner of Police, Nyeri Area to Commissioner of Police, Nairobi, 'Stock Thefts – Mweiga/Ngobit – Trackers', 30 July 1954, National Archives, UK, WO 276/248. For a mention of Becker in Mau Mau see G. Sinclair, *At the End of the Line: Colonial Policing and the Imperial Endgame* (Manchester: Manchester University Press, 2006), p. 160.
74. 'Anti-Mau Mau Operations', p. 12.
75. Hiscox, *Dawn Stand-to*, p. 286.
76. P. Hewitt, *Kenya Cowboy: A Police Officer's Account of the Mau Mau Emergency* (Johannesburg: 30 Degrees South, 2008), pp. 228–42, 252–4.
77. D. Franklin, *A Pied Cloak: Memoirs of a Colonial Police (Special Branch) Officer* (London: Janus Publishing, 1996), pp. 99, 55.
78. Bartlett, *Shoot Straight*, pp. 79–80.
79. L. Smith, Interview with T. G. Symons, Imperial War Museum, 2007, reel 8.
80. W. Itote (General China), *Mau Mau General* (Nairobi: East African Publishing House, 1967), p. 73.
81. 'Training for Forest Operations', Kenya Regiment Training Pamphlet, 1953, p. 12, obtained from J. Davis.
82. Hespeler-Boultbee, *Mrs. Queen's Chump*, p. 93.
83. M. Clough, *Mau Mau Memoirs: History, Memory and Politics* (Boulder, CO: Lynne Rienner Publishers, 1998), p. 151.
84. 'Training for Forest Operations', Kenya Regiment Training Pamphlet, 1953, p. 12.
85. Hewitt, *Kenya Cowboy*, pp. 109–10.
86. H. Miller, 'Electronic Devices in Malaya's Jungle War', *The Straits Times*, 9 March 1958, p. 9. I was led to this article by L. Comber, *Malaya's Secret Police, 1945–60: The Role of the Special Branch in the Malayan Emergency* (Victoria, Australia: Monash University Press, 2008), p. 103.
87. C. Pugsley, *From Emergency to Confrontation: The New Zealand Armed Forces in Malaya and Borneo, 1949–66* (Oxford: Oxford University Press, 2003), pp. 144–5; and W. D. Baker, *Dare to Win: The Story of the New Zealand Special Air Service* (Melbourne: Lothian, 1987), p. 29.
88. For stereotypes about the Maori see H. Beattie, 'The Maori as a Tracker and Signaller', *The*

New Zealand Railways Magazine, 14:5 (August 1939), pp. 40–4. For Woods see P. Bush and P. Thomas, *Peter Bush: A Life in Focus* (Auckland: Hodder Moa, 2009), p. 143.

89. Field Marshal J. Harding to Secretary of State for Colonies and Kenya Governor, 'Trackers', 1 June 1956, National Archives, UK, FO 371/123893; Sir E. Baring to Secretary of State for Colonies, 7 June 1956, National Archives, UK, FO 371/123896; and M. Blundell, *So Rough A Wind: The Kenya Memoirs of Sir Michael Blundell* (London: Weidenfeld and Nicholson, 1964). For the dogs see B. Grob-Fitzgibbon, *Imperial Endgame: Britain's Dirty Wars and the End of Empire* (London: Palgrave-MacMillan, 2011), p. 334. For Turkish trackers in Kenya see C. Yennaris, *From the East; Conflict and Partition in Cyprus* (London: Elliot and Thomson, 2003), pp. 129–30.

90. F. Kitson, *Bunch of Five* (London: Faber and Faber, 1977), pp. 183–91; J. Peterson, *Oman's Insurgencies: The Sultanate's Struggle for Supremacy* (London: Saqi, 2007), p. 172; D. Smiley and P. Kemp, *Arabian Assignment* (London: Leo Cooper, 1975), p. 97; and F. Kitson, *Low Intensity Operations: Subversion, Insurgency and Peacekeeping* (London: Faber and Faber, 1971), p. 136.

91. Kitson, *Low Intensity Operations*, p. 193.

92. Steinhart, *Black Poachers, White Hunters*, p. 196.

93. Parker, *Colonial Regiment*, p. 277.

94. 'Noel Simon', Obituary, *The Telegraph*, 30 October 2008; and Parker, *Colonial Regiment*, p. 216.

95. Holman, *Elephants at Sundown*, pp. 84–115.

96. Steinhart, *Black Poachers, White Hunters*, p. 201; and 'Peter Jenkins', Obituary, *The Telegraph*, 13 October 2001.

97. Parker, *Last Colonial Regiment*, pp. 284–7, 290, 309–11.

98. Steinhart, *Black Poachers, White Hunters*, p. 198.

99. 'Jack Barrah', Obituary, *The Telegraph*, 9 August 2013; and Parker, *Colonial Regiment*, p. 267.

100. 'Stanley (Stan) Richard Bleazard', *Old Cambrian Society*, at http://www.oldcambrians.com/Alumni-Bleazard,Stanley.html [accessed 15 April 2015].

101. I. Parker, 'Tributes: Rodney Elliot (1921–2005)', *Swara Magazine*, 29 (2006), p. 55.

4 Rhodesia (Zimbabwe), 1965–80

1. There is a particularly large amount of literature on the Rhodesian conflict. For some important examples see D. Martin and P. Johnson, *The Struggle for Zimbabwe: The Chimurenga War* (New York: Monthly Review, 1981); J. K. Cilliers, *Counter-Insurgency in Rhodesia* (Beckenham: Croom Helm Publishing, 1985); T. Ranger, *Peasant Consciousness and Guerrilla War in Zimbabwe: A Comparative Study* (London: James Currey, 1985); T. Ranger and N. Bhebhe (eds), *Soldiers in Zimbabwe's Liberation War* (London: James Currey, 1995); P. Moorcroft and P. McLaughlin, *The Rhodesian War: A Military History* (London: Pen and Sword, 2008); J. Nhongo-Simbanegavi, *For Better or Worse: Women and ZANLA in Zimbabwe's Liberation Struggle* (Harare: Weaver Press, 2000); and E. Sibanda, *The Zimbabwe African People's Union, 1961–1987* (Trenton: Africa World Press, 2005).

2. D. Kennedy, *Islands of White: Settler Society and Culture in Kenya and Southern Rhodesia, 1890–1930* (Durham, NC: Duke University Press, 1987); J. Bonello, 'The Development of Early Settler Identity in Southern Rhodesia, 1890–1914', *International Journal of African Historical Studies*, 43:2 (2010), pp. 341–67; T. Stapleton, *No Insignificant Part: The Rhodesia Native Regiment and the East African Campaign of the*

First World War (Waterloo: Wilfrid Laurier University Press, 2006); and T. Stapleton, *African Police and Soldiers in Colonial Zimbabwe, 1923–80* (Rochester: University of Rochester Press, 2011).

3. B. Cole, *The Elite: The Story of the Rhodesian Special Air Service* (Amazimtoti, South Africa: Three Knights Publishing, 1984), p. 11.

4. T. G. Turner-Dauncey, 'The Happy Hundred'; and L. M. Coetzee, 'Lessons from Malaya', in J. Pittaway and C. Fourie (eds), *SAS Rhodesia* (Musgrave, South Africa: Dandy Agencies, 2003), pp. 58, 88.

5. Interview with O. Veremu, Vumba, Mutare, 18 July 2009. For accounts of the RAR in Malaya see A. Binda, *Masodja: The History of the Rhodesian African Rifles and its Forerunner the Rhodesia Native Regiment* commissioned and compiled by Brigadier David Heppenstall and the Rhodeisan African Rifles Regimental Association, UK (Johannesburg: 30 Degrees South, 2007), pp. 126–40; and J. Essex-Clark, *Maverick Soldier: An Infantryman's Story* (Melbourne: Melbourne University Press, 1991), pp. 3–15, 32–49.

6. Interview with P. Mufanebadza, Gutu, 5 August 2009.

7. W. A. Godwin, quoted in Binda, *Masodja*, p. 12.

8. Essex-Clark, *Maverick Soldier*, p. 36.

9. L. Charlton, 'Defence of the Nation', *Horizon*, 8 (1966), n.p.

10. R. Reid-Daly, 'War in Rhodesia: Cross-Border Operations', in A. J. Venter (ed.), *Challenge: Southern Africa Within the African Revolutionary Context* (Gibraltar: Ashanti Publishing, 1989), p. 156.

11. D. Schafer, *Simply the Greatest Life: Finding Myself in the Country* (Bloomington, IN: Balboa Books, 2012), p. 196.

12. Communication with A. Savory, 11 January 2013; Interview with B. Robinson (former commanding officer of the Rhodesian SAS), Durban, 24 June 2014; Lieutenant Colonel B. Robinson, 'Into the Bush War', in Pittaway and Fourie, *SAS*, p. 94; Reid-Daly, 'War in Rhodesia', pp. 157–8; and statement by D. Watt, https://www.facebook.com/124280565850/posts/10152194012215851 [accessed 15 August 2014].

13. P. French, *To the Edge: Shadows of a Forgotten Past with the Rhodesian SAS and Selous Scouts* (Solihull: Helion and Company, 2012), pp. 38, 46.

14. J. R. D. Wood, *Zambezi Valley Insurgency: Early Rhodesian Bush War Operations* (Solihull: Helion and Company, 2012), p. 20.

15. T. Bopela and D. Luthuli, *Umkhonto we Sizwe: Fighting for a Divided People* (Alberton: Galago, 2005), p. 63.

16. Binda, *Masodja*, pp. 206–7.

17. D. Scott-Donelan, 'Zambezi Valley Manhunt', *Soldier of Fortune*, March 1985. For the insurgent infiltrations of Rhodesia in 1967 see R. M. Ralinala, J. Sithole, G. Houston and B. Magubane, 'The Wankie and Sipolilo Campaigns', *The Road to Democracy in South Africa (1960–70)*, (Cape Town: Zebra Press, 2004), vol. 1, pp. 479–540.

18. B. Orford, *Kamchacha: Rhodesian Game Ranger* (Bulawayo, Zimbabwe: privately published, 2008), pp. 227–8.

19. P. Gibbs, H. Phillips and N. Russell, *Blue and Old Gold: The History of the British South Africa Police, 1889–1980* (Johannesburg: 30 Degrees South, 2009), pp. 334–43; and Stapleton, *African Police and Soldiers*. For PATU using African game-trackers in 1968 see P. J. H. Petter-Bowyer, *Winds of Destruction: The Autobiography of a Rhodesian Combat Pilot* (Johannesburg: 30 Degrees South, 2005), p. 133. For the enrolment of African trackers in the police reserve see Communication with Anthony Trethowan, 10 April 2013.

20. Rhodesian Ministry of Information and Tourism, *Rhodesian Commentary* (Salisbury: Rhodesian Government, 1967), vols 3–5, p. 11. See also Gibbs, Phillips and Russell, *Blue and Old Gold*, p. 342.

21. A. J. Venter and friends, *War Stories: Up Close and Personal in Third World Conflicts* (Pretoria: Protea Book House, 2011), p. 96.

22. D. Lemon, *Never Quite a Soldier: A Rhodesian Policeman's War 1971–82* (Stroud: Albita Books, 2000), p. 123.

23. A. Trethowan, *Delta Scout: Ground Coverage Operator* (Johannesburg: 30 Degrees South, 2008), pp. 129, 135–6.

24. D. H. Hubbard Jr, *Bound for Africa: Cold War Fight Along the Zambezi* (Annapolis, MD: Naval Institute Press, 2008), pp. 79, 184, 220.

25. Moorcroft and McLaughlin, *The Rhodesian War*, p. 56.

26. For BSAP tracker dogs see J. R. T. Wood, *A Matter of Weeks Rather Than Months: The Impasse Between Harold Wilson and Ian Smith, Sanctions, Aborted Settlements and war, 1965–69* (Bloomington, IN: Trafford Publishers, 2012), p. 363, 373, 437; and Petter-Bowyer, *Winds of Destruction*, pp. 126, 139–40.

27. K. Thomas, *Shadows in an African Twilight* (Cape Town: Uthekwane Press, 2008), pp. 115–17; Communication with Anthony Trethowan, 10 April 2012; and Orford, *Kamchacha*, pp. 228–9.

28. Thomas, *African Twilight*, p. 117.

29. Thomas, *African Twilight*, pp. 118–20.

30. Communication with Trethowan; Nick Tredger, *From Rhodesia to Mugabe's Zimbabwe: Chronicles of a Game Ranger* (Alberton: Galago, 1999), p. 28; and Orford, *Kamchacha*, p. 229.

31. R. Selley, *West of the Moon: Early Zululand and a Game Ranger at War in Rhodesia* (Johannesburg: 30 Degrees South, 2009), p. 213.

32. Thomas, *African Twilight*, p. 139.

33. D. Croukamp, *The Bush War in Rhodesia: The Extraordinary Combat Memoir of a Rhodesian Reconnaissance Specialist* (Boulder, CO: Paladin Press, 2006), p. 88.

34. Croukamp, *Bush War*, p. 89.

35. Wood, *A Matter of Weeks*, p. 447.

36. Robinson, 'Tracker Combat Teams', in Pittaway and Fourie, *SAS*, p. 204.

37. For the first tracking course see Croukamp, *Bush War*, pp. 80, 88. For the creation of the Tracking Wing in 1970 see Wood, *A Matter of Weeks*, p. 373; Robinson, 'Tracker Combat Teams', p. 205; and Interview with Robinson.

38. Interview with Robinson.

39. Thomas, *African Twilight*, p. 122.

40. K. Thomas, 'Night of the Lioness', *Hunters Network*, 29 October 2009, at http://www.africahunting.com/threads/night-of-the-lioness.14782/ [accessed 17 April 2015]. See also Thomas, *African Twilight*, pp. 89–98; and Interview with Robinson.

41. T. Bax, *Three Sips of Gin: Dominating the Battlespace with Rhodesia's Elite Selous Scouts* (Solihull: Helion and Company, 2013), p. 169.

42. Binda, *Masodja*, p. 315; and F. Watts, 'Death of Al Tourle', in M. Adams and C. Cocks, *Africa's Commandos: The Rhodesian Light Infantry* (Solihull: Helion, 2013), p. 95.

43. Bax, *Three Sips of Gin*, p. 131.

44. Petter-Bowyer, *Winds of Destruction*, 140–1; A. Binda, *The Saints: The Rhodesian Light Infantry*, ed. C. Cocks (Johannesburg: 30 Degrees South, 2007), p. 80; and Reid-Daly, 'War in Rhodesia', pp. 160–1.

45. J. P. Cann, *The Flechas: Insurgent Hunting in Eastern Angola, 1965–1974* (Pinetown, South Africa: 30 Degrees South, 2013); R. K. Hitchcock, 'Refugees, Resettlement, and Land and Resource Conflicts: The Politics of Identity Among! Xun and Khwe San in Northeastern Namibia', *African Study Monographs*, 33:2 (June 2012), pp. 73–132, on p. 83.

46. S. Nielsen, 'The White Devil of Mozambique: One Man Army Against FRELIMO', *Soldier of Fortune*, October 1978, pp. 78–83; and P. Nortje, *32 Battalion: The Inside Story of South Africa's Elite Fighting Unit* (Cape Town: Zebra Press, 2003), p. 86. After Mozambique's independence in 1974, Roxo joined the South African Special Forces and was attached to the SADF's 32 Battalion. He participated in the 1975 South African invasion of Angola where he was killed in a landmine incident the next year.

47. P. E. S. Correia, 'Political Relations Between Portugal and South Africa from the End of the Second World War until 1974' (PhD Thesis, University of Johannesburg, 2007), p. 210.

48. A. J. Venter, *Portugal's Guerrilla Wars in Africa: Lisbon's Three Wars in Angola, Mozambique and Portuguese Guinea 1961–74* (Solihull: Helion and Company, 2013), pp. 379–90.

49. For Portuguese troops at the Tracking Wing see A. Binda, 'Mozambique, 1968–1972: Rhodesian and Portuguese Cooperation', *Rhodesian Forces*, at www.rhodesianforces. org/Mozambique1968–72.htm [accessed 17 April 2015]. For Portuguese officers in Rhodesia and tracking demonstrations see Croukamp, *Bush War*, pp. 133–4. For Angola see Robinson, 'Tracker Combat Teams', p. 205; and Interview with Robinson.

50. Cole, *The Elite*, p. 38.

51. Binda, *The Saints*, pp. 104–10; Croukamp, *Bush War*, pp. 135–9; and Venter, *Portugal's Guerrilla Wars*, pp. 387–8.

52. D. Price, 'RLI Tracking', in Adams and Cocks, *Africa's Commandos*, p. 56.

53. Price, 'RLI Tracking', pp. 56–8.

54. Binda, *The Saints*, p. 130; and *The Cheetah*, 31 October 1980, pp. 44–5.

55. Price, 'RLI Tracking', p. 58.

56. Communication with Vince Leonard and Malcolm Clewer, 2 August 2013.

57. A. Binda, *The Rhodesia Regiment: From Boer War to Bush War, 1899–1980* (Alberton, South Africa: Galago, 2012), pp. 190, 415.

58. Binda, *Masodja*, p. 285 mentions the use of PATU trackers by RAR and Game Department trackers by RLI in 1976.

59. For ZAPU in 1968 see Binda, *Saints*, p. 94. For 1970 see 'Rhodesian Air Force Anti-terrorist Operations (COINOPS), Early COIN Ops in Rhodesia', *Rhodesian Forces*, at www.rhodesianforces.org/AntiTerrorismOps.htm [accessed 17 April 2015]. For ZANU in 1968 see Wood, *A Matter of Weeks*, p. 497.

60. Moorcroft and McLaughlin, *The Rhodesian War*, p. 100; and Thomas, *African Twilight*, p. 121.

61. A. Mutambara, *The Rebel in Me: A ZANLA Guerrilla Commander in the Rhodesian Bush War, 1975–1980* (Pinetown, South Africa: 30 Degrees South, 2014), pp. 64–8.

62. Mutambara, *Rebel in Me*, pp. 78–9.

63. Reid-Daly, 'War in Rhodesia', p. 164.

64. Trethowan, *Delta Scout*, p. 131.

65. Cole, *The Elite*, p. 38.

66. Binda, *Rhodesia Regiment*, p. 339.

67. H. Ellert, *Rhodesian Front War: Counter-Insurgency and Guerrilla War in Rhodesia, 1962–80* (Gweru: Mambo Press, 1989), pp. 27–8. For more on the cordon see Cilliers,

Counter-Insurgency in Rhodesia, p. 107.

68. Ellert, *Rhodesian Front War*, p. 28.
69. Cilliers, *Rhodesian Counter-Insurgency*, pp. 15, 83.
70. J. Parker, *Assignment Selous Scouts: Inside Story of a Rhodesian Special Branch Officer* (Alberton: Galago, 1999), p. 113.
71. R. R. Daly (as told to P. Stiff), *Selous Scouts: Top Secret War* (Alberton: Galago, 1982), pp. 82–3.
72. Major General Archer Bruce Campling, DCD, 'Pseudo-Terrorist Operations in Rhodesia', 2006, supplied by Kevin Thomas.
73. For recruiting of trackers from other units see Robinson, 'Into the Bush War', p. 195; Price, 'RLI Tracking', p. 58; and Parker, *Assignment*, p. 31. For poaching see Croukamp, *Bush War*, p. 383; Thomas, *Shadows*, pp. 278–81; and S. Ellis, 'Of Elephants and Men: Politics and Nature Conservation in South Africa', *Journal of Southern African Studies*, 20:1 (March 1994), pp. 55–6.
74. Parker, *Assignment Selous Scouts*, p. 56.
75. J. R. D. Wood (ed.) *The War Diaries of Andre Dennison* (Gibraltar: Ashanti Publishing, 1989), pp. 95–6, 175.
76. E. Bird, *Special Branch War: Slaughter in the Rhodesian Bush Southern Matabeleland, 1976–1980* (Solihull: Helion, 2014), p. 75.
77. Bird, *Special Branch War*, p. 123.
78. Bird, *Special Branch War*, p. 233. Other examples are listed on pp. 58, 79, 86, 90, 119, 177, 201, 229.
79. Binda, *Rhodesia Regiment*, pp. 278–9.
80. Interview with M. Watson, *Soldier of Fortune*, 22 December 2012.
81. Communication with Anthony Trethowan, 21 June 2013; C. Reynolds, 'Forces Bag 31 in One Battle', *Rhodesia Herald*, 18 November 1976, pp. 1–2; '31 Terrs Killed in "Bloody Fine Effort"', *Rhodesian Forces*, at http://www.rhodesianforces.org/31Terrorists.htm [accessed 17 April 2015]; Petter-Bowyer, *Winds of Destruction*, p. 280.
82. Binda, *Saints*, p. 200. Laurie Ryan was killed in a hunting accident after the war.
83. Binda, *Rhodesia Regiment*, p. 415.
84. Binda, *Rhodesia Regiment*, p. 278; and French, *To The Edge*, pp. 73–6.
85. Binda, *Rhodesia Regiment*, pp. 323, 326, 338.
86. B. Whyte, *A Pride of Eagles: The Story of Rhodesia's Air Force* (Salisbury: Graham Publishing, 1976), p. 43.
87. Whyte, *Pride of Eagles*, pp. 43, 55.
88. T. Grundy and B. Miller, *The Farmer at War* (Salisbury: Modern Farming Publications, 1979), p. 29.
89. Ellert, *Rhodesian Front War*, p. 91 mentions SAP combat tracker team in Matibi TTL in 1979.
90. J. Lott, 'Run the Bastards Down: C.A.T.U. Tracks Terrorists; Rhodesia's Civilian Tracking Unit', *Soldier of Fortune Magazine*, July 1979, pp. 46–51. For Dunn see Binda, *Rhodesia Regiment*, p. 251;
91. Bird, *Special Branch War*, pp. 11, 79, 86, 100, 116, 173.
92. Parker, *Selous Scouts*, pp. 252–3.
93. Petter-Bowyer, *Winds of Destruction*, pp. 325–6
94. Wood, *Dennison*, p. 141.
95. Croukamp, *Bush War*, pp. 380–1.
96. Wood, *Dennison*, p. 140.

97. Trethowan, *Delta Scout*, p. 159.
98. Binda, *Rhodesia Regiment*, p. 332.
99. Binda, *Rhodesia Regiment*, p. 340.
100. Parker, *Selous Scouts*, p. 136
101. Bax, *Three Sips*, p. 245.
102. P. McAleese, *No Mean Soldier: The Story of the Ultimate Professional Soldier in the SAS and Other Forces* (London: Orion, 1994), p. 153.
103. Bax, *Three Sips*, p. 260.
104. Bird, *Special Branch War*, pp. 213–4.
105. For external operations see Croukamp, *Bush War*, pp. 226, 244, 250, 262, 308, 343, 366, 379. For FRELIMO see J. Greeff, *A Greater Share of Honour: The Memoirs of a Recce Officer* (Durban: Just Done, 2008), p. 99. For Bushmen see M. Bolaane, 'The Role Played by Botswana During the Liberation Struggle', Hashim Mbita Project.
106. J. Alexander, 'Dissident Perspectives on Zimbabwe's Post Independence War', *Africa: Journal of the International Africa Institute*, 8:2 (1998), pp. 151–82; and 'ZNA Looking for GI Janes', *The Zimbabwean*, 18 October 2011.
107. K. Vorlaufer, 'CAMPFIRE – The Political Ecology of Poverty Alleviation, Wildlife Utilization and Biodiversity Conservation in Zimbabwe', *Erdkunde*, 56:2 (April–June 2002), pp. 184–206.
108. G. H. Tatham, 'The Rhino Conservation Strategy in the Zambezi Valley Codenamed Operation Stronghold', *The Zimbabwe Science News*, 22:1–2 (January–February 1988), p. 21.
109. B. Chilvers, 'Rhino's Last Stand in Africa', *REF Journal*, 3 (1990), p. 17.
110. M. Vollers, 'The Rhino Wars: Zimbabwe is Shooting Poachers Who Menace the Rare Black Rhino', *Sports Illustrated*, 2 March 1987; T. Masland, 'Zimbabwe Losing War to Save Rhino', *Chicago Tribune*, 27 January 1989; and R. Duffy, 'The Role and Limitations of State Coercion: Anti-Poaching Policies in Zimbabwe', *Journal of Contemporary African Studies*, 17:1 (1999), pp. 97–120.
111. K. Meadows, *Sometimes When It Rains: White Africans in Black Africa* (Thorntree Press, 2000), pp. 299–300; and S. Barnes, 'The Fall and Rise of the Rhino', *The Times*, 31 January 2009.
112. P. Clemence and B. Clemence, 'Aggressive Tracking Training End of Year Report', *Save Foundation Newsletter*, December 2011, pp. 2–3, emphasis is original. For Clemence in the Rhodesian Army see Thomas, *African Twilight*, pp. 278–9; and Binda, *The Saints*, p. 108.
113. Communication with Head of Anti-Poaching at a conservancy in Zimbabwe.
114. Communication with Trethovan, 21 June 2013, quoting Binda.

5 South West Africa (Namibia), 1966–90

1. For the German colonial period see H. Bley, *South West Africa under German Rule, 1894–1914* (Evanston, IL: Northwestern University Press, 1971); H. Dreschler, *Let Us Die Fighting: The Struggle of the Herero and Nama Against German Imperialism (1884–1915)* (London: Zed Press, 1980); J. Gewald, *Herero Heroes: A Socio-Political History of the Herero of Namibia, 1890–1923* (Oxford: James Currey, 1999); and J. Sarkin, *Germany's Genocide of the Herero: Kaiser Wilhelm II, His General, His Settlers, His Soldiers* (Woodbridge: James Currey, 2011). For the South African invasion during the First World War see B. Nasson, *Springboks on the Somme: South Africa and the Great War, 1914–1918* (Johannesburg: Penguin, 2007), pp. 63–87; and H. Strachan,

The First World War in Africa (Oxford: Oxford University Press, 2004), pp. 61–92. For the liberation struggle see P. H. Katjavivi, *A History of Resistance in Namibia* (London: James Currey, 1988); E. I. Udogu, *Liberation Namibia: The Long Diplomatic Struggle Between the United Nations and South Africa* (Jefferson, CO: Macfarland, 2012); and C. Leys and J. Saul, *Namibia's Liberation Struggle: The Two Edged Sword* (London: James Currey, 1995). For a general overview see M. Wallace, *A History of Namibia: From the Beginning to 1990* (New York: Columbia University Press, 2011).

2.   S. Brown, 'Diplomacy by Other Means – SWAPO's Liberation War', in Leys and Saul (eds), *Nambia's Liberation Struggle*, pp. 19–29; C. J. Nothling, 'Military Chronicle of South West Africa (1915–1988)', *South African Defence Force Review* (1989); and A. Clayton, *Frontiersmen: Warfare in Africa Since 1950* (London: Routledge, 1999), pp. 117–18.

3.   There is a large literature on the South West African/Angolan War. For example, see E. George, *The Cuban Intervention in Angola, 1965–1991: From Che Guevara to Cuito Cuanavale* (London: Frank Cass, 2005); P. Gleijeses, *Conflicting Missions: Havana, Washington and Africa, 1959–1976* (Chapel Hill, NC: University of North Carolina Press, 2002); M. Norval, *Death in the Desert: The Namibian Tragedy* (Washington D.C.: Selous Foundation, 1989); E. G. M. Alexander, 'The Cassinga Raid' (MA Thesis, University of South Africa, 2003); J. W. Turner, *Continent Ablaze: The Insurgency Wars in Africa 1960 to the Present* (Johannesburg: Jonathan Ball, 1998), pp. 69–84; W. Steenkamp, *South Africa's Border War, 1966–1989* (Gibraltar: Ashanti, 1989); W. Steenkamp, *Borderstrike! South Africa into Angola, 1975–80* (Durban: Just Done Productions, 2006); D. Williams, *On the Border: The White South African Military Experience, 1965–1990* (Cape Town: Tafelberg, 2008); P. Polack, *The Last Hot Battle of the Cold War: South Africa vs. Cuba in the Angolan Civil War* (Oxford: Casemate, 2013); and L. Scholtz, *The SADF in the Border War, 1966–1989* (Cape Town: Tafelberg, 2013).

4.   A. Seegers, 'One State, Three Faces; Policing in South Africa (1910–1990)', *Social Dynamics*, 17:1 (1991), pp. 36–48; A. Grundlingh, '"Protectors and Friends of the People?" The South African Constabulary in the Transvaal and Orange River Colony, 1900–1908', in D. M. Anderson and D. Killingray (eds), *Policing the Empire: Government, Authority and Control, 1830–1940* (Manchester: Manchester University Press, 1991), pp. 168–82; M. Dippenaar, *The History of the South African Police, 1912–1988* (Pretoria: *Promedia*, 1988); and H. Hamann, *Days of the Generals: The Untold Story of South Africa's Apartheid Era Military Generals* (Cape Town: Zebra Press, 2001), p. 9.

5.   A. Seegers, *The Military in the Making of Modern South Africa* (London: I. B. Tauris, 1996), pp. 92–110; A. Peled, *A Question of Loyalty: Military Manpower Policy in Multi-Ethnic States* (Ithaca, NY: Cornell University Press, 1998), pp. 27–125; B. Sass, 'The Union and South African Defence Force, 1912 to 1994', in J. Cilliers and M. Reichardt (eds), *About Turn: The Transformation of the South African Military and Intelligence* (Halfway House, South Africa: Institute for Defence Policy, 1995), pp. 118–39; R. S. Jaster, 'South African Defense Strategy and the Growing Influence of the Military', in W. Foltz and H. Bienen (eds), *Arms and the African: Military Influences on Africa's International Relations* (New Haven, CT: Yale University Press, 1985), pp. 125–6; W. Minter, *Apartheid's Contras: An Inquiry into the Roots of War in Angola and Mozambique* (London: Zed Books, 1994), p. 122; and G. Cawthra, *Brutal Force: The Apartheid War Machine* (London: International Defence and Aid Fund for Southern Africa, 1986), pp. 89–110.

6.   L. Marshall, 'Marriage Among !Kung Bushmen', *Africa: Journal of the International Africa Institute*, 29:4 (October 1959), pp. 335–65, on p. 362.

7. *The African Communist*, 28 (First Quarter 1967), p. 10.
8. K. Shear, 'Police Dogs and State Rationality in Early Twentieth Century South Africa', in L. V. Sittert and S. Swart (eds), *Canis Africanis: A Dog History of Southern Africa* (Leiden: Brill, 2008), pp. 193–216.
9. L. J. Bothma, *Buffalo Battalion: South Africa's 32 Battalion: A Tale of Sacrifice* (Bloemfontein: L. J. Bothma, 2007), p. 31.
10. R. Lee and S. Hurlich, 'From Foragers to Fighters: South Africa's Militarization of the Namibian San', in E. Leacock and R. Lee (eds), *Politics and History in Band Societies* (Cambridge: Cambridge University Press, 1982), pp. 334–5, 340–1; and M. Bolaane, 'The Role Played by Botswana During the Liberation Struggle', unpublished paper, Hashim Mbita Project, Dar-es-Salaam, Tanzania. In the 1980s the government of Botswana passed a law prohibiting its citizens from enlisting in foreign security forces.
11. H. MacMillan, *The Lusaka Years: The ANC in Exile in Zambia, 1963–1994* (Auckland Park, South Africa: Jacana Media, 2013), p. 36.
12. D. Franklin, *A Pied Cloak: Memoirs of a Colonial Police (Special Branch) Officer* (London: Janus Publishing, 1996), p. 156.
13. T. J. Stapleton, Interview with Francois du Toit, 14 November 2012, Radium; T. J. Stapleton, Interview with Herman van der Walt, 15 November 2012, Polokwane; T. J. Stapleton, Interview with Kallie Calitz, 16 November 2012, Odendaalsrus; G. Cawthra, *Brutal Force: The Apartheid War Machine* (London: International Defence and Aid Fund, 1986), pp. 123, 225; and 'Learning to Survive in a Life-or-Death Situation', *Paratus* (May 1981), pp. 30–1.
14. P. Stiff, *The Covert War: Koevoet Operations Namibia, 1979–89* (Alberton, South Africa: Galago, 2004), pp. 47–50.
15. S. Kamongo and L. Bezuidenhout, *Shadows in the Sand: A Koevoet Tracker's Story of an Insurgency War* (Johannesburg: 30 Degrees South Publishers, 2011), pp. 37–41; T. J. Stapleton, Interview with S. Kamongo, 14 November 2012, Springbok Flats, Warmbad; and T. J. Stapleton, Communication with G. Manning, 20 January 2012.
16. *Truth and Reconciliation Commission of South Africa Report, Volume 2* (Cape Town: Truth Reconciliation Commission, 1998), p. 1. For a broad discussion see P. E. S. Correia, 'Political Relations Between Portugal and South Africa from the End of the Second World War until 1974' (PhD Thesis, University of Johannesburg, 2007), pp. 141–51.
17. Colonel J. Breytenbach, cited in R. Reeve and S. Ellis, 'An Insider's Account of the South African Security Forces' Role in the Ivory Trade', *Journal of Contemporary African Studies*, 13:2 (1995), pp. 227–44.
18. K. Grundy, *Soldiers Without Politics: Blacks in the South African Armed Forces* (Berkeley, CA: University of California Press, 1983), p. 254; P. Nortje, *32 Battalion: The Inside Story of South Africa's Elite Fighting Unit* (Cape Town: Zebra Press, 2003), p. 17; and I. Uys, *Bushman Soldiers: Their Alpha and Omega* (Germiston, South Africa: Fortress Publications, 1993), pp. 1–25.
19. R. Gordon and S. S. Douglas, *The Bushman Myth: The Making of a Namibian Underclass* (Boulder, CO: Westview Press, 2000), pp. 1–2, 183–5. For a more celebratory account see Uys, *Bushman Soldiers*; G. B. Kolata, '!Kung Bushmen Join the South African Army', *Science*, 211:4482 (February 1981), pp. 562–4; and A. K. Battistoni and J. J. Taylor, 'Indigenous Identities and Military Frontiers: Reflections on San and the Military in Namibia and Angola, 1960–2000', *Lusotopie*, 16:1 (2009), pp. 113–31. For the introduction of mixed-race people, Asians and blacks into the SADF see A. Peled, *A Question of Loyalty: Military Manpower Policy in Multiethnic States* (Ithaca, NY and

London: Cornell University Press, 1999), pp. 27–92. For evidence of forced recruitment see testimony of Agostinho Victorino, 29 October 1998 in *Truth and Reconciliation Commission of South Africa Report, Volume 2* (Cape Town: Truth Reconciliation Commission, 1998), p. 23.

20. C. Enloe, *Ethnic Soldiers: State Security in Divided Societies* (Athens, GA: University of Georgia Press, 1982).

21. R. Gordon, 'People of the Great White Lie?', *Cultural Survival Quarterly*, 15:1 (Spring 1991), at https://www.culturalsurvival.org/ourpublications/csq/article/people-great-white-lie [accessed 18 April 2015].

22. S. Brown, 'Diplomacy by Other Means – SWAPO's Liberation War', in C. Leys and J. Saul (eds), *Namibia's Liberation Struggle: The Two-Edged Sword* (London: James Currey, 1995), pp. 29–39.

23. Cawthra, *Brutal Force*, p. 181.

24. J. H. Thompson, *An Unpopular War: From Afkak to Bosbefok: Voices of South African National Servicemen* (Cape Town: Zebra Press, 2006), p. 118.

25. A. Esterhuyse and E. Jordaan, 'The South African Defence Force and Counterinsurgency, 1966–1990', in D. Baker and E. Jordaan (eds), *South Africa and Contemporary Counterinsurgency; Roots, Practises and Prospects* (Claremount, South Africa: International Publishers, 2010), pp. 111–13.

26. N. Basson and B. Motinga, *Call Them Spies* (Windhoek, Namibia: Africa Communications Project, 1989), p. 12.

27. H. R. Heitman, *South African War Machine* (Novato, CA: Presidio Press, 1985), p. 157.

28. Kamongo and Bezuidenhout, *Shadows in the Sand*, p. 47.

29. Uys, *Bushman Soldiers*, pp. 49, 54–6.

30. Uys, *Bushman Soldiers*, p. 144. For more information from Uys, *Bushman Soldiers*, see: p. 49 for training by the Recces; p. 56 for Bushmen transferred to the Recces; pp. 156–7 for tracking courses; and pp. 64–6, 109 for 31 Battalion Recce Wing. For another discussion of Bushmen in the Recces, see P. J. Els, *We Fear Naught But God* (South Africa: privately published, 2009), p. 143.

31. Gordon and Douglas, *Bushman Myth*, p. 186.

32. J. Cock and L. Bernstein, *Melting Pots and Rainbow Nations: Conversations about Difference in the United States and South Africa* (Chicago, IL: University of Illinois Press, 2002), p. 100.

33. O. O. Namakalu, *Armed Liberation Struggle: Some Accounts of PLAN's Combat Operations* (Windhoek, Namibia: Gamsberg MacMillan, 2004), p. 137.

34. T. J. Stapleton, Interview with P. Ekandjo, 21 July 2014, Windhoek, Namibia.

35. Gordon, 'People of the Great White Lie?'.

36. *Paratus*, April 1978, p. 28.

37. *Paratus,* May 1974, p. 53.

38. M. Norval, 'SADF's Bushman Battalion: Desert Nomads on SWAPO's Track', *Soldier of Fortune*, March 1984, p. 74.

39. '5 SAI (1969)', *Sentinel Projects*, 1 July 2000, at http://sadf.sentinelprojects.com/bg1/5sai69.html [accessed 18 April 2015].

40. A. Feinstein, *In Conflict* (Windhoek, Namibia: New Namibia Books, 1998), p. 38.

41. T. Ramsden, *Border-line Insanity: A National Serviceman's Story* (Alberton, South Africa: Galago Books, 1999), p. 125.

42. A. M. Behr, 'Etosha Area Force Unit', *Paratus*, May 1986, pp. 14–15; and U. Dieckmann, *Haillom in the Etosha Region: A History of Colonial Settlement, Ethnicity and*

Nature Conservation (Basel, Switzerland: Basler Afrika, 2007), p. 206.

43. Behr, 'Etosha Area Force Unit', p. 15.
44. 'Our Dog School', *Paratus*, May 1971, p. 61; and 'Dog Centre is Here to Stay', *Paratus*, July 1989, p. 15.
45. Sittert and Swart (eds), *Canis Africanis*, p. 29.
46. P. Stark, *The White Bushman* (Pretoria: Protea Book House, 2011), p. 56.
47. J. J. P. de Vries and S. Swart, 'The South African Defence Force and Horse Mounted Infantry Operations, 1974–1985', *Scientia Militaria*, 40:3 (2012), pp. 398–428. For the Portuguese see J. P. Cann, *Counterinsurgency in Africa: The Portuguese Way of War, 1961–1974* (Solihull: Helion and Company, 2012), pp. 140–5; A. M. Behr, '1 SWA Specialist Unit – Otavi', *Paratus*, May 1986, pp. 10–12; H.-R. Heitman, *South African Armed Forces* (Cape Town: Buffalo Productions, 1990), pp. 109–11; Uys, *Bushmen Soldiers*, p. 157; J. Geldenhuys, *At the Front: A General's Account of South Africa's Border War* (Johannesburg: Jonathan Balls, 2009), pp. 105–106; H.-R. Heitman, *South African War Machine* (Novato, CA: Presidio Press, 1985), pp. 108–110; and Namakalu, *Armed Liberation Struggle*, p. 139.
48. Stapleton, Interview with Ekandjo.
49. Kamongo and Bezuidenhout, *Shadows in the Sand*, pp. 62–3.
50. P. Stiff, *The Covert War: Koevoet Operations Namibia, 1979–89* (Alberton, South Africa: Galago, 2004), pp. 50–67; and S. Kamongo and L. Bezuidenhout, *Shadows in the Sand: A Koevoet Tracker's Story of an Insurgency War* (Johannesburg: 30 Degrees South Publishers, 2011), pp. 26–7.
51. J. Hooper, *Koevoet: Experiencing South Africa's Deadly Bush War* (1988; Warwickshire: GG Books, 2012), p. 116.
52. Stiff, *Covert War*, pp. 67–73.
53. E. de Kock, *A Long Night's Damage: Working for the Apartheid State* (Johannesburg: Contra Press, 1998), p. 77.
54. P. Stark, *The White Bushman* (Pretoria: Protea Book House, 2011), p. 206.
55. Stiff, *Covert War*, p. 109.
56. A. Durand, *Zulu Zulu Golf: Life and Death with Koevoet* (Cape Town: Zebra Press, 2011), p. 227.
57. A. Durand, *Zulu Zulu Foxtrot: To Hell and Back with Koevoet* (Cape Town: Zebra Press, 2012), pp. 116–17.
58. Durand, *Zulu Zulu Golf*, p. 144; T. J. Stapleton, Communication with G. Manning, 20 January 2012; and T. J. Stapleton, Interview with S. Kamongo, 14 November 2012, Springbok Flats, Warmbad.
59. Kamongo and Bezuidenhout, *Shadows in the Sand*, pp. 26–7, 68, 70, 138–9.
60. Durand, *Zulu Zulu Golf*, for dogs see pp. 233, 237; and for environment see p. 250.
61. H. Grobler as told to L. Bezyidenhout, 'Courage, Blood and Treachery', in Kamongo and Bezuidenhout, *Shadows in the Sand*, pp. 245–63, on p. 249; and Stiff, *Covert War*, p. 267.
62. J. Greeff, *A Greater Share of Honour: The Memoirs of a Recce Officer* (Durban: June Done, 2008), pp. 138–9.
63. Kamongo and Bezuidenhout, *Shadows in the Sand*, p. 27.
64. *Truth and Reconciliation Commission of South Africa Report, Volume 2* (Cape Town: Truth Reconciliation Commission, 1998), pp. 72–7.
65. H. Hamann, *Days of the Generals: The Untold Story of South Africa's Apartheid Era Military Generals* (Cape Town: Zebra Press, 2001), p. 65. Similar views were strongly expressed by generals Jannie Geldenhuys and Georg Meiring.

66. D. Lord, *From Fledgling to Eagle: The South African Air Force During the Border War* (Johannesburg: 30 Degrees South Publishers, 2008), p. 251.
67. Uys, *Bushman Soldiers*, pp. 103, 117. While Uys claims that the RM system was pioneered by 201 Battalion and then adopted by the rest of the SADF, it is clear that it was first employed by Koevoet in the late 1970s.
68. '101 Battalion's Romeo Mikes Hot on the Heals of SWAPO Terrorists', *Paratus*, October 1985, pp. 16–18; 'Men Who Keep the Romeo Mikes Geared for Action', *Paratus*, December 1985, pp. 24–5; P. V. D. Merve, 'Colour for 101 Battalion: Fighting for True Freedom in SWA', *Paratus*, December 1988, p. 31; W. Steenkamp, 'Politics of Power – The Border War', in A. J. Venter (ed.), *Challenge: Southern African Within the African Revolutionary Context – An Overview* (Gibraltar: Ashanti Publishing, 1989), pp. 207–23; and Stiff, *Koevoet Operations*, pp. 220–1, 248–9, 254.
69. *The South African Defence Force Yearbook*, 1987, p. 305; and *Report of the Truth and Reconciliation Commission*, 29 October 1998, vol. 2, p. 23.
70. Reuters TV, 18 July 1985, at http://www.itnsource.com/shotlist//RTV/1985/07/23/BGY601220242/?s=* [accessed 18 April 2015].
71. Korff, *Bullet*, p. 196.
72. D. Kirkman, '7 SAI: Memories of My SADF Experiences', 1 July 2000, at http://sadf.sentinelprojects.com/kirk/7sai.html [accessed 18 April 2015].
73. 'Operational Medical Orderly and Operational Planning Officer, 1985–1996', 1 July 2000, at http://sadf.sentinelprojects.com/bg1/7opsmed.html [accessed 18 April 2015].
74. Korff, *Bullet*, p. 127.
75. Thompson, *An Unpopular War*, p. 121.
76. Esterhuyse and Jordaan, 'Counterinsurgency', p. 113; Scholtz, *Border War*, p. 198; G. Korff, *19With a Bullet: A South African Paratrooper in Angola* (Johannesburg: 30 Degrees South, 2009), p. 129; and M. Alexander, 'South African Airborne Operations', *Scientia Militaria: South African Journal of Military Studies*, 31:1 (2001), pp. 49–82.
77. Lord, *From Fledgling to Eagle*, p. 252.
78. Interview with L. Musore, 14 November 2012, Warmbads.
79. A. J. Venter, *Gunship Ace: The Wars of Neall Ellis, Helicopter Pilot and Mercenary* (Newbury: Casemate, 2011), pp. 100–5. For teams carrying helicopter fuel see K. Hooper, *Koevoet*, p. 138 and for cooperation with helicopters see quotation by G. Manning, pp. 148–9.
80. Venter, *Gunship Ace*, p. 105.
81. Kamongo and Bezuidenhout, *Shadows in the Sand*, p. 48.
82. Lord, *From Fledgling to Eagle*, pp. 245–7.
83. Interview with P. Ekandjo, Windhoek, 21 July 2014.
84. Korff, *Bullet,* p. 127.
85. M. Paul, *Parabat: Personal Accounts of Paratroopers in Combat Situations in South Africa's History* (Benoryn: My Books Productions, 2008), p. 77.
86. Steenkamp, *South Africa's Border War*, p. 295.
87. P. Ekandjo, *The Volunteers' Army, 1962–1989* (Windhoek: Privately Published, 2014), pp. 26–58; P. Ekandjo, *The Jungle Fighter* (Windhoek: Privately Published, 2011), pp. 36–64; Interview with Ekandjo; Namakalu, *Armed Liberation Struggle*, p. 156; Communication with Gavin Manning (former Koevoet), 31 January 2012, Since many Koevoet members were former insurgents, they knew a great deal about PLAN training methods.
88. D. Herbstein and J. Evenson, *The Devils Are Among Us: The War for Namibia* (London: Zed Press, 1989), p. 75.
89. Ekandjo, *Jungle Fighter*, p. 297.

90. Interview with S. Kamongo, 14 November 2012, Springbok Flats, Warmbad; Interview with Francois du Toit, 14 November 2012, Warmbad; Interview with Callie Kalitz, 16 November 2012, Odendaalsrus; Communication with Gavin Manning, 31 January 2012; and Korff, *19 With a Bullet*, pp. 126–35. For the 1971 incident see A. Vines, *Still Killing: Landmines in Southern Africa* (New York: Human Rights Watch Arms Project, 1997), p. 104.

91. Kamongo and Bezuidenhout, *Shadows in the Sand,* p. 150.

92. Durand, *Zulu Zulu Foxtrot*, p. 17.

93. Kamongo and Bezuidenhout, *Shadows in the Sand*, pp. 120–1, 133–4, 251.

94. Namakalu, *Armed Liberation Struggle*, pp. 134–6.

95. SWAPO Department of Information and Publicity, *To Be Born a Nation: The Liberation Struggle for Namibia* (London: Zed Press, 1981), p. 172.

96. Thompson, *Unpopular War*, p. 195.

97. J. Breytenbach, *The Buffalo Soldiers: The Story of South Africa's 32-Battalion, 1975–1993* (Alberton: Galago, 1999), pp. 188, 190, 197, 206–7. See also Bothma, *Buffalo Battalion*; and Nortje, *32 Battalion*.

98. *Korff, 19 With a Bullet*, p. 184.

99. Greeff, *A Greater Share of Honour*, p. 56. For training see P. Stiff, *The Silent War; South Africa Recce Operations*, 1969–1994 (Alberton: Galago, 2006), pp. 38, 448.

100. Els, *We Fear Naught But God*, pp. 58, 131; and H. McCallion (Henry Gow*)*, *Killing Zone: A Life in the Paras, the Recces, the SAS and the RUC* (London: Bloomsbury, 1996), pp. 89–91.

101. Stiff, *Silent War*, p. 448.

102. Greeff, *A Greater Share of Honour*, pp. 132, 136–7.

103. R. Williams, 'The Other Armies: A Brief Historical Overview of Umkhonto we Sizwe (MK) 1961–1994', *South African Military History Journal*, 11:5 (June 2000); S. Ellis, *External Mission: The ANC in Exile, 1960–90* (Oxford University Press, 2013); K. M. Kondlo, '"In the Twilight of the Azanian Revolution": The Exile History of the Pan Africanist Congress of Azania (South Africa) (1960–1990)' (PhD Thesis, Rand Afrikaans University, 2003). For pepper see H. Barrell, Interview with R. Kasrils, September 1990, Johannesburg, at http://www.nelsonmandela.org/omalley/cis/omalley/OMalleyWeb/03l v03445/04lv04015/05lv04154/06lv04159/07lv04173.htm [accessed 18 April 2015]. For ANC military training in the 1970s see ANC Reports in Training in East Africa, South Africa Political Materials, University of the Witwatersrand, A2675, Part 3, Folder 22.

104. For police tracking see Interview with V. D. Walt. For MK deaths see 'Justice Minister to Hand Over Exhumed Remains of Former MK Members', at http://www.justice.gov.za/m_statements/2014/2014–03–05-mk-remains.html [accessed 18 April 2015]; and 'Diary of MK Operative', William Cullen Library, University of the Witwatersrand, A3149A.

105. J. McMullin, *Ex-Combatants and the Post-Conflict State: Challenges of Reintegration* (Palgrave-MacMillan, 2013), pp. 89–90; and Ekandjo, *Jungle Fighter*, pp. 79, 103.

106. Greef, *Greater Share*, xii; Interview with Corporal P. Ndinda (Field Ranger), 9 November 2012, Skukuza, Kruger National Park; M. Cadman, Interview with M. P. Schofield, 21 August 2007, Project Missing Voices, University of the Witwatersrand. For Greef's company see http://www.ntomenirangers.com/ [accessed 18 April 2015].

107. A. Grossman, 'Lost COIN: The Erosion of Counterinsurgency in the South African Army', in Baker and Jordaan, *South Africa and Contemporary Counterinsurgency*, p. 142.

108. M. Appel, 'SA's Silent Guerrilla War', *The New Age*, 17 April 2013, at http://www.thenew-age.co.za/mobi/Detail.aspx?NewsID=92378&CatID=1007 [accessed 18 April 2015].

109. G. Reeder, *Middelburg Observer*, 15 April 2013; and M. Haufiku, 'Police Complete Probe in Human Trafficking Case', *New Era*, 20 September 2013.
110. D. Henk, *The Botswana Defence Force in the Struggle for an African Environment* (New York: Palgrave-MacMillan, 2007), pp. 53–61; Interview with former BDF soldier M. Bannink, 15 November 2012, Polokwane; and Sergeant T. King, 'U.S. Botswana Special Forces Train Together', at http://www.army.mil/article/83250/U_S___Botswana_Special_Forces_train_together/ [accessed 18 April 2015].
111. Durand, *Zulu Zulu Foxtrot*, p. 17.

Conclusion

1.   K. Agger, 'Field Dispatch: Chasing the Lord's Resistance Army', Enough Project, November 2012, at http://www.enoughproject.org/files/ChasingTheLRA.pdf [accessed 18 April 2015].
2.   E. Brandon, 'White House Sending Ospreys to Hunt Joseph Kony and the LRA', *Christian Science Monitor*, 26 March 2014; J. Vandiver, 'Airlift Support Gets Pulled Out of Effort to Counter Lord's Army', *Stars and Stripes*, 10 April 2014; and 'African Troops Capture Junior LRA Rebel Commander', *Fox News*, 22 April 2014. See also L. Cline, *The Lord's Resistance Army* (Santa Barbara, CA: Praeger Security International, 2013).

INDEX

101 Specialist Unit, SADF, 117

Aberdares Mountains, Kenya, 44
Adamson Force, 49
Adamson, George, Kenya Game Department, 49, 53, 56, 66
African National Congress (ANC), South Africa, 71, 79, 102, 110, 131–3
Aggressive Tracking Specialists (ATS), Zimbabwe, 98
Algeria, 4, 6, 108
Amatola Mountains, 34
Ambambi, "Communist", SWAPO officer, 129
American Civil War, 28
Anderson, Major G. H., hunter, 24
Anglo-Boer War
    First (1880–1), 36
    Second (1899–1902), 17, 36–9, 106–7, 116
Anglo-Ndebele War (1893–4), 35
Angola, 7, 17, 40, 70, 85, 101–7, 109–13, 116, 118–22, 127–8, 130–3, 135, 139–41
anti-poaching campaigns, 142
Anti-tracking, 78, 95
Australia, 28–9, 37, 41, 43, 98, 107

Baden-Powell, R. S. S., British officer, 35–8, 41, 75, 99, 137
Bagshawe, F. J., British District Officer, 24
Bailey, Bill, British South Africa Police, 80
Bakalahadi, 24
Baldwin, William Charles, hunter, 14
Bantustans, South Africa, 108
Baring, Sir Evelyn, Governor of Kenya, 44
Barrah, Jack, Kenya Game Department, 67
Bartlett, Fred, Kenya Game Department, 49, 63
Basarwa, 13

Battalions
    7 South African Infantry Battalion (7 SAI), SADF, 123
    21 Battalion, SADF, 122
    31 Battalion, SADF (later 201 Battalion, SWATF), 111, 113, 115, 117, 121–2
    32 Battalion, SADF, 105, 111, 119, 130–1
Bax, Tim, Rhodesian Officer, 84, 95
Bechuanaland, 17, 19, 40, 61, 72, 102, 110
Becker, Peter, hunter, 62
Belgian Congo, 22–3
Biesele, Megan, anthropologist, 18
Bird, Ed, Rhodesian Special Branch, 91
Bleazard, Stan, Kenya Regiment, 55, 66–7
Boers (Afrikaners), 13, 29–30, 32–3, 36–7
Bonham, Captain Jack, Kenya Game Department, 62
Bopela, Thula, African National Congress, 79
Bothma, Steve, Kenya Regiment, 56
Botswana, 13, 16–17, 19, 24, 27, 61, 72–3, 86–7, 89, 93, 95–6, 102, 109–10, 131–2, 134
Botswana Defence Force (BDF), 134
Bousfield, Don, Kenya Game Department, 53, 63
Boy Scouts, 38
Bridger, Jim, American scout, 28
Brigades
    39 Brigade, British Army, Kenya, 44, 46
    49 Brigade, British Army, Kenya, 44, 58
    70 Brigade, British Army, Kenya, 44, 55
Brind, Peter, British officer, 55
British South Africa Company (BSAC), 33, 35, 69
British South Africa Police (BSAP), 71, 80–1, 83, 89, 91, 95

Bromwich, Mike, Rhodesian Wildlife Officer, 82
Brown, Monty, Kenya Game Department, 51, 53
Bryden, H. Anderson, hunter, 16–17
Bumppo, Natty, fictional character, 28
Burbridge, Ben, hunter, 23–4
Burma, 5, 51, 53, 76
Burnham, Frederick Russell, American tracker, 17, 33, 36, 38, 41, 75, 99
Bushmanland Borderers, 37
Bushmen, 11–19, 24–5, 29, 33, 40–1, 61, 80–1, 85, 96, 99, 109–17, 119–20, 123–4, 130, 134, 137, 139–40
Bwabwata National Park, Namibia, 134

Cabora Bassa, Mozambique, 72
Callwell, Charles E. British officer and writer, 37
Campbell, Alex, safari industry pioneer, 24
Campbell, Guy, Kenya Regiment, 50, 52
Cape Colony, 29–30
Cape Corps, South Africa, 31, 108
Cape Mounted Rifles (CMR), 13, 29–31, 34, 41
Cape-Xhosa Wars, 107
Caprivi Strip, Namibia, 17, 103–4, 109, 111, 113–14, 130, 134
Capstick, Peter Hathaway, hunter and writer, 19
Carson, Kit, American scout, 28
Cassinga, Angola, 105
Central African Federation, 70, 77
Central African Republic (CAR), 23–4, 142–4
Central Kalahari Game Reserve (CKGR), 19
Chimoio, Mozambique, 74
China, 46, 64, 102
Chitepo, Herbert, ZANU leader, 73
Christy, Cuthbert, explorer, 22
Citizen Force, South Africa, 105, 108
Civilian African Tracking Unit (CATU), Rhodesia, 93–4
Clark, John Desmond, archaeologist, 18
Clemence, Pete, Rhodesian Army, 86, 98
Cobra Teams, South African Police, 110
Cocks, Chris, Rhodesian Army, 92
Coetzee, Pinkie, SADF, 113
Cold War, 3–4, 7, 45, 104
Cole, Berkeley, British aristocrat, 38

Colenbrander, Johannes, Boer tracker, 33
Combat Tracking Special Groups, Portuguese Army Mozambique, 86
Conway, Joe, Rhodesian Army, 79
Cooke, Graham, game ranger South Africa, 21
Corbett, Jim, Lieutenant Colonel, British Army, 53
Counter-insurgency, 140
Cowie, Colonel Melvin, Kenya National Parks Department, 62
Croukamp, Denis, Rhodesian Army, 82
Cuba, 132
Cuito Cuanavale, Angola, 107
Cyprus, 65

De Beer, Dewald, SADF, 130
De Beer, Willie, Rhodesian Army , 86
De Gentile, Geoffroy, hunter, 23
De Kock, Eugene, South African Police, 118–19, 140
Democratic Republic of Congo (DRC), 22, 24, 104, 133, 142–4
Dennison, Andre, Major, Rhodesian Officer, 91
Department of National Parks and Wildlife Management (DNPWLM), Rhodesia, 81
De Wit, Chris, South African Police, 119
dog school
    SADF, Bourke's Luck, 116
    South African Police, Potgietersrus, 109–10
Domestic animals and counter-insurgency, 140
Dornan, Samuel Shaw, missionary, 17
Dreyer, Hans, Colonel, South African Police, 118–19, 140

East Africa Association, 44
East Africa Battle School, British Army, 57, 141
East Africa Command, British Military, 53, 65
East African Professional Hunters' Association, 25
Eastern Bloc, 3, 45, 70, 132
Ekandjo, Peter, PLAN, 114, 118, 128
Ellert, Henrik, Rhodesian Intelligence, 90
Elliott, Rodney, Kenya Game Department, 48, 53, 57–9, 66–7, 137
Embu, 46, 52, 67
Equestrian Centre, SADF, Potchefstroom, 116–17
Erskine, Francis, Kenya Regiment, 55–6

Erskine, George, General, British Army, 45
Essex-Clarke, John, Rhodesian and Australian officer, 76–7
Etosha Area Defence Unit, SWATF, 115
Etosha National Park, Namibia, 103, 115–17, 119, 133, 140

Feinstein, Anthony, SADF, 115
Ferreira, T. Captain, SWATF, 122
Fey, Venn, Kenya Regiment, 53, 58–60, 68, 138
Fingo (Mfengu), 30–1, 34–5
Fire Force, 91–2, 94, 125
Folliott, Captain R.J. British officer, 61
Foran, William Robert, hunter, 20
Forest Operating Company, Kenya, 59–60
Forestry Department, Kenya, 49
Fort Hall, Kenya, 49, 58
Fort Jericho, Kenya, 58
Fort Peddie, Cape Colony, 32, 34
Franklin, Derek, Kenya Police, 63, 110
Free Fire Zones, 141
French, Paul, Rhodesian Army, 79
Front for the Liberation of Mozambique (FRELIMO), 85–6, 88, 96, 132
Frontier Armed and Mounted Police (FAMP), 30, 107

Game Department
  Kenya, 48–9, 51–3, 62, 67
  Northern Rhodesia, 77
  Southern Rhodesia, 77–8
Gaomab, Johannes, PLAN, 112
Gatti, Attilio, Italian explorer, 23
Gaza, Mozambique, 96
Gichimu, Ndorobo tracker, 52–3, 58, 138
Gill, Leonard, Kenya Regiment, 53
Gillmore, Parker, hunter, 16, 24
Giriama, 51
Gloucestershire Regiment, 61
Godbeer, Ray, SADF and Rhodesian Army, 131
Godwin, W.A. Rhodesian officer, 76
Gonarezhou National Park, Zimbabwe, 82, 98
Gordon-Cumming, Roualeyn, hunter, 13
Graaf Reinet, South Africa, 29
Grahamstown, South Africa, 32
Grasselli, Giorgio, hunter, 80

Greef, Jack, SADF, 133–4
Grey's Scouts, Rhodesian Army, 92–3, 116
Grootboom, Jan, Fingo tracker, 35
Grootfontein, South West Africa, 115
Ground Coverage, British South Africa Police, 80, 89
Guerrilla Anti-Terrorist Unit (GATU), Rhodesia, 78, 80, 83
Guyu, Galo-Galo, Kenyan tracker, 49, 51

Hadza, 2, 24
Hainyeko, Tobias, PLAN commander, 104, 128
Harding, John, British Governor of Cyprus, 65
Hatton, Denis Finch, hunter, 20, 38
Herero, 40, 101
Hespeler-Boultbee, J.J., British officer, 55
Hickman, John, Rhodesian officer, 75
Holmes, Sherlock, fictional character, 38
Holub, Emil, Czech explorer, 16
Home Guard, Kenya, 52, 58
Honde Valley, Zimbabwe, 92
Hughes, Robin, Rhodesian Wildlife Officer, 12, 82
Hunter, John Alexander, hunter, 24

Imperial British East Africa Company, 43
Inkomo Barracks, Rhodesia, 94
Intelligence Force (I Force), Kenya, 50
Intensive Protection Zones, Zimbabwe, 98
Ionides, C.J.P., game warden Tanganyika, 25
Isiolo, Kenya, 49, 66
Itote, Waruhiu (General China), Mau Mau commander, 45

Jenkins, Peter, Kenya National Parks Department, 53, 67
Jungle Warfare School (JWS), British Army, 5, 65, 76, 138

Kalahari Desert, 2, 13, 72, 101–2
Kalanga, 73
Kamba, 25, 39, 51, 65–8, 138
Kamino, Kenya Game Scout, 50
Kamongo, Sisingi, Koevoet member, 126
Kaokoland, 103, 105–6, 118, 121–2
Kariba, Rhodesia, 70, 72, 79, 82–4, 86–7, 90, 93
Kasciula, Twa chief and tracker, 23

Kavango, 103, 105–6, 113, 118, 121, 127
Kazungula, 110
Kearney, Dennis, Kenya Game Department, 67
Kelly, Ned, Australian outlaw, 29
Kenya African Union (KAU), 44
Kenya Defence Force (KDF), 43
Kenya Land Freedom Army, 6–7, 44–64, 66–8, 78, 110, 137–8, 141–4
Kenya Police, 44, 48, 50–1, 53, 56, 58, 62–3, 67, 142
Kenya Police Reserve (KPR), 48, 51, 65–6
Kenya Regiment, 43–4, 48–56, 58, 61–2, 65–7
Kenyatta, Jomo, Kenyan Nationalist leader, 44, 47
Khoikhoi, 13, 17, 29–32, 34, 41
Khoisan, 13, 27, 114
Kiambu, Kenya, 56, 58
Kikuyu, 7, 43–4, 46–8, 50–3, 55–6, 58, 61, 63, 68, 138
Kilonzo, Kibwezi, Kenya Game Department, 51
Kimathi, Dedan, Mau Mau commander, 45, 47, 61
King's African Rifles (KAR), 43–4, 49–52, 54–5, 59–61, 64, 66
Kipsigi, 52–3, 68
Kirkman, Derek, SADF, 123
Kitchener, Lord Horatio, British officer, 37
Kitson, Frank, British officer, 5, 55, 66–7
Kleinboy, guide, 14
Koevoet, 118–23, 125–9, 135, 140, 142, 144
Korean War, 52
Korff, Granger, SADF, 123, 127, 130
Kruger, J. Major, SWATF, 123
Kruger National Park, South Africa, 21, 110, 133–4
Kwanyama Ovambo, 101

Lancashire Fusiliers, 67
Land Apportionment Act, Rhodesia, 69
Lathbury, Sir Gerald, British officer, 46, 61
League of Nations, 101
Lee, Richard, anthropologist, 18
Legion of Frontiersmen, 38, 40
Leninism, 3
Limpopo River, 72
Livingstone, David, 13
Livingstone Museum, 18

Lobengula, Ndebele King, 16, 33
Londolozi Game Reserve, South Africa, 21
Loots, Fritz, Major General, SADF, 113
Lord, Dick, Brigadier, SAAF, 121
Lord's Resistance Army (LRA), 142–4
Lord Strathcona's Horse, 37
Luanda, Angola, 104, 112
Lusaka Agreement (1984), 106
Luthuli, Daluxola, African National Congress, 79, 81
Luthuli Detachment (1967), 79, 81

Maasai, 11, 24, 52, 63, 67
MacArthur, Duncan, Rhodesian Army, 89, 95
Makuleke, 21
Malawi, 67, 70
Malaya, 4–5, 46, 56–7, 64–5, 73, 75–7, 84, 87, 90, 138
Maleoskop, South African Police Counter-insurgency School, 110
Mana Pools National Park, Zimbabwe, 82
Manyangadze, Kenneth, Zimbabwe National Parks, 98
Maoism, 3
Mao Zedung, 3, 73
Maplanka, Bushman tracker, 95
Marshall, John, film maker, 18
Marshall, Lorna, ethnographer, 18
Mashonaland, Zimbabwe, 81
Matabeleland, Zimbabwe, 73, 79, 95–6
Mathebula, Carlson, tracker, 21
Mathenge, Stanley, Mau Mau commander, 45
Mau Mau, 7, 45–50, 52, 54–6, 58, 60–1, 63–4, 66, 68, 138, 141
Maun, Botswana, 134
Mavinga, Angola, 105
McCabe, David, Kenya Game Department, 67
McKinnon, George, British official, 31
Mears, Ray, survival expert, 25
Meinertzhagen, Richard, Major, British officer, 39
Meru, 46, 52
Military Areas
    1 Military Area (1MA), South West Africa, 105
    2 Military Area (2MA), South West Africa, 105
Mills, Peter, Kenya Police, 56, 63

Mount Kenya, 44–5, 48, 53, 57, 63, 68, 138
Movement for the Popular Liberation of Angola (MPLA), 102, 104–5, 112, 132
Mozambique, 7, 11, 21, 70, 72–4, 82, 84–9, 92, 95–6, 119, 131–3, 139, 141
Mpondo, 11, 27
Mufanebadza, Paul, Rhodesian Army, 76
Mugabe, Robert, Zimbabwe Nationalist Leader, 70, 74–5
Musore, Laurens, Koevoet member, 126
Mutambara, Agrippah, ZANLA insurgent, 88
Mweiga, Kenya, 62–3

Nairobi, Kenya, 44–6, 63
Nama, 101
Namaqualand Border Scouts, 37
Namib Desert, 103
Namibia, 6–7, 9, 13, 107, 133, 135, 137
Nandi, 20, 38–9, 43, 52, 66
Nanyuki, Kenya, 53–4, 62
Natal, 25, 32–3, 133
Natal Parks Board, 25, 113, 130
National Front for the Liberation of Angola (FNLA), 104–5, 111, 119
Nationalist Party, South Africa, 102, 108
National Parks Department
    Kenya, 48, 62, 66
    Southern Rhodesia, 86
    Zimbabwe, 97–8
Native Mounted Police, Queensland, Australia, 28
Ndebele, 11, 14, 16, 27, 33, 35–6, 69, 73, 75
Ndemanga, Aly, Malawian gun-bearer and tracker, 12
Ndlambe, Xhosa leader, 29
Ndorobo, 11, 20, 25, 50, 52–3, 58, 63, 68, 137–8
Nel, Chris, South African Police, 110, 119
New Zealand military, 65
Ngalu, Kenyan tracker, 54
Ngobit, Kenya, 62–3
Ngonyo, Kiribai, Sergeant, Kenya Regiment, 51, 66
Ngqika, Xhosa leader, 29
Ngwato, 14, 27
Niassa, Mozambique, 85
Nieuwenhuizen, Piet, Boer tracker, 40
Nkomo, Joshua, Zimbabwe Nationalist Leader, 70, 74

Northern Rhodesia, 17–18, 24, 69–71, 76
North-West Mounted Police, Canada, 37–8
Ntombela, Magqubu, game scout South Africa, 25
Nujoma, Sam, SWAPO leader, 102
Nyadzonya, Mozambique, 74
Nyamahanga, Warrant Officer, KAR, 55, 64
Nyasaland, 70–1, 76–7

Ohangwena, South West Africa, 110
Oman, 65–7
Omega Camp, South West Africa, 121
Ondangwa, South West Africa, 122
Ongulumbashe, South West Africa, 103
Ongwediva, South West Africa, 119
Operation
    Anvil (1954), Kenya, 46, 55
    Askari (1983), Angola, 106
    Cauldron (1968), Rhodesia, 82
    Gordion Knot (1970), Mozambique, 72
    Hammer (1955), Kenya, 59
    Hurricane, Rhodesia, 82
    Jock Scott (1952), Kenya, 44
    Modular (1987), Angola, 107
    Protea (1981), Angola, 106
    Reindeer (1978), Angola, 105
    Savannah (1975), Angola, 104, 111, 113
    Save Our Heritage, Zimbabwe, 97
    Sceptic (1980), Angola, 106, 127
    Stronghold, Zimbabwe, 97–8
Orange River Colony Volunteers, 37
Organization of African Unity (OAU), 71, 74, 102
Oshakati, South West Africa, 105, 118
Oshikango, South West Africa, 104
Oshivelo, South West Africa, 117
Otavi, South West Africa, 117
Ovambo, 25, 101, 110–12, 118–20, 122–3, 135, 139–40, 142
Ovambo Home Guard, 118, 122
Ovambo Home Guard , 119
Ovamboland, 103, 105–6, 111–13, 115–16, 118–22, 127–8

Pan-Africanist Congress (PAC), South Africa, 131–2
Parker, Ian, Kenya Regiment, 48, 67
People's Liberation Army of Namibia (PLAN), 102–7, 110, 112–14, 118–22, 126–33

Percival, A. Blayney, Kenya Game Warden, 21

persistence hunting, 1, 140

Petter-Bowyer, Peter, Rhodesian Air Force, 84, 93

Pocock, Roger, Founder, Legion of Frontiersmen, 38

Police Anti-Terrorist Unit (PATU), Rhodesia, 80, 82, 87, 138, 142

Police Reserve Air Wing (PRAW), Rhodesia, 81

Portugal, 7, 70, 72, 102, 104, 111

Portuguese, 7–8, 72–4, 84–6, 89, 101–3, 111, 116, 131–2, 139–41

Potgietersrus, South Africa, 110

Potts, Gordon L., Major, British Army, 52, 54

Powell-Cotton, P. H. G., hunter, 22

Pretorius, P. J., Major, Boer tracker, 22, 39

Price, Don, Rhodesian Officer, 86, 93

pygmies, 22–3

Ragati Forest Station, Kenya, 50

Rainsford, William Stephen, hunter, 20

Ramsden, Tim, SADF, 115

Reid Daly, Ron, Rhodesian officer, 74, 85–6, 88, 90

Reitz, Deneys, 36

Rhodes, Cecil, 33, 69

Rhodesia, 7–9, 19, 21, 25, 40, 56, 70–5, 77–9, 85–90, 92–3, 96, 103, 106, 113, 116, 118–19, 128, 131, 137, 139–42

Rhodesian African Rifles (RAR), 71, 74–6, 79, 84, 87, 90–1, 99

Rhodesian Air Force, 81, 84, 93, 139

Rhodesian Front, 70, 74

Rhodesian Light Infantry (RLI), 71, 74, 79, 82, 84, 86–7, 90, 92, 96, 131, 142

Rhodesia Regiment (RR), 71, 87, 89, 91–5

Robinson, Brian, Rhodesian officer, 83, 86

Roosevelt, Theodore, 20

Roxo, Daniel, Portuguese hunter, 85–6

Royal Air Force (RAF), 46

Royal Green Jackets, British Army, 60

Royal Inniskilling Fusiliers, British Army, 49

Royal Irish Fusiliers, British Army, 61

Royal Northumberland Fusiliers, British Army, 59

Rundu, South West Africa, 105

Ruyter, tracker, 13

Rwanda, 23

Ryan, Laurie, Rhodesian Army, 87, 92

Samburu, 50, 52, 66

San, 13, 18, 114

Saul, Sergeant, Bushman tracker Namibia, 40

Save Valley Conservancy, Zimbabwe, 98

Savory, Allan, Rhodesian officer, 77–8, 80, 83

Scammel, Dave, Wildlife Officer, Rhodesia, 81

sectors

   Sector South West Africa, 10, 106, 122

   Sector South West Africa, 20, 106

   Sector South West Africa, 70, 106

Selley, Ron, Rhodesian Wildlife Officer, 82

Selous, Frederick Courtney, hunter, 16–17, 20, 40, 75, 99

Selous Scouts, 74, 87, 90–6, 98–9, 118, 130–1, 139, 142

Shangaan, 11, 21–2, 25, 92–4, 96, 99, 137, 139

Shangani Patrol, 33

Shaw, William, missionary, 31

Sheldrick, Daphne, Kenya conservationist, 56

Sheldrick, David, Kenya National Parks Department, 66–7

Shona, 11, 69, 72–3, 75

Simba, Hekuta, Kenyan tracker, 49, 51, 66

Sim, Jack, Kenya Game Department, 48

Simon, Noel, Kenya Wildlife Society and National Parks Department, 66

Sithole, Ndabaningi, Zimbabwe Nationalist Leader, 70, 74

Skeepers, C. J. Rhodesian Army, 84

Smiley, Colonel David, British officer, 65

Smit, Gysberd, Rhodesian tracker, 80

Smith, Harry, British governor/Noel, Lieutenant, Rhodesian Officer, 31

Smith, Ian, Rhodesian Premier, 70–1, 74, 78

Smith, Noel, Lieutenant, Rhodesian Officer, 91

Smithsonian Institute, 20

Social Darwinism, 15

Somali, 49, 66

Somerset, Colonel Henry, British officer, 29, 32

South Africa, 6–8, 11, 13, 16, 21, 30, 37–9, 69–72, 74, 101–5, 109–10, 118, 121–2, 127, 131–4, 139–41

South African Air Force (SAAF), 121, 125–6, 140

South African Constabulary, 37, 107
South African Defence Force (SADF), 104, 106, 108–18, 121–2, 125, 127, 129–30, 134–5, 140
South African Mounted Rifles, 108
South African National Defence Force (SANDF), 134
South African Police (SAP), 50, 93, 107, 109–10, 113, 118, 132, 139
Southern Rhodesia, 6, 16–17, 33, 37, 49, 53, 69–71, 77
South West Africa, 6–9, 19, 25, 40, 101–7, 109–14, 116, 121–2, 127–8, 130–1, 133–5, 137, 139–42, 144
South West African National Union (SWANU), 102
South West African People Union (SWAPO), 102–7, 109–13, 118, 120–2, 124, 127–31, 135, 140
South West African Specialist Unit (SWASPES), 117–18, 142
South West African Territorial Force (SWATF), 106, 115, 117, 122
Soviet Union, 3, 102
Soweto Uprising (1976), South Africa, 132
Sparrow, Mark, Rhodesian Army, 93–4
Special Air Service, 5, 65, 71, 75–9, 83, 86–90, 92, 95–6, 99, 138, 142
    Britain, 83
    New Zealand, 5, 65
    Rhodesia, 79, 131
Special Branch, 57, 59, 61
    Botswana, 110
    Rhodesia, 91, 95
Special Field Force, Namibian Police, 133
Special Forces, 2, 71, 79, 86, 96, 109, 131
    South Africa, 84, 105, 111, 113, 118–19, 121, 130, 133
    United States, 135, 142
Special Task Force, South African Police, 110
Stark, Peter, SADF, 116, 119, 140
Steele, Nick, game warden South Africa, 25
Steele, Sam, Canadian officer, 37
Stevenson-Hamilton, James, chief warden Kruger National Park, 21
Stigand, Captain C.H., hunter, 20
Stockil, Clive, Zimbabwe conservationist, 21
Stock Owners Association of Kenya, 62

Super-ZAPU, 96
Support Unit, British South Africa Police and Zimbabwe Republic Police, 80, 96
Swazi, 21
Swaziland, 131, 133

Tanganyika, 12, 24–5, 44, 55, 102
Tanzania, 2, 72, 86, 102, 131
Tatham, Glen, Zimbabwe National Parks, 97
Taylor, John, hunter, 12
Tete, Mozambique, 72–3, 85–6
Thomas, Elizabeth Marshall, ethnographer, 18
Tiroyamodimo, Otisitswe B. BDF, 134
Tobias Hainyeko Training Centre, Lubango, Angola, 128
Tobias, Phillip, anthropologist, 18
Tongogara, Josiah, ZANLA commander, 73
Tooley, Jim, Kenya Regiment, 53, 61
Tourle, Al, Rhodesian Officer, 84
Tracker Combat Team (TCT), Kenya, 57, 59–60, 62, 138
Tracker Combat Unit (TCU), Rhodesia, 78–83, 90, 99, 138, 142
Tracker Unit, SADF, 113
Tracking School, Kenya, 53–4, 57, 62, 137–8, 141
Tracking Wing
    Kenya, 58
    Rhodesian Army, 83, 85, 87, 90, 93, 98, 139, 142
Transvaal National Scouts, 37, 41
Transvaal Scottish, SADF, 114
Trethowan, Anthony, British South Africa Police, 89, 95
Tribal Trust Lands, Rhodesia, 69
Tsandi, South West Africa, 115, 118
Tsavo National Park, Kenya, 51, 67
Tshuma, Nelson, Namibian tracker, 95
Tsumeb, South West Africa, 115
Tswana, 13–14, 31
Turkana, 50–2, 65–6
Twa, 11, 22–4

Ulundi, Battle of (1879), 32
Umkhonto we Sizwe (Spear of the Nation or MK), South Africa, 132–3
Unilateral Declaration of Independence (UDI), Rhodesia 70–1, 77, 1965

Union Defence Force (UDF), 107–8
Union for the Total Independence of Angola (UNITA), 102–5, 107, 111–12, 121, 127–8, 132
United Nations (UN), 6, 102, 107
United States Agency for International Development (USAID), 97
United States Army, 5, 28

Van Der Post, Laurens, writer and film maker, 18
Van Deventer, Tom, SADF, 113
Veremu, Obert, Rhodesian Army, 76
Vietnam, 4–5, 65, 76, 108, 126
Viljoen, Constand, General, SADF, 121
Virunga Mountains, 23
Voi Field Force, Kenya, 66–7, 142
Volunteer Tracking Unit (VTU), Rhodesia, 81–2
Von Lettow-Vorbeck, Paul, German officer, 24, 38, 40

Waata (Liangulu), Kenya, 49, 51
Walls, Peter, Rhodesian general, 75, 94
Wankie National Park, Rhodesia/Zimbabwe, 80, 83, 86–7
War Council, Britain, 61–2
Waterkloof Highlands, 34
Watt, Darrell, Rhodesian army, 78
Welensky, Roy, Central African Federation Prime Minister, 77
White Highlands, Kenya, 44
Wienholt, Arnold, Australian scout, 17, 40
Wilson, Major Allan, BSAC, 33

Woodley, Bill, Kenya National Parks Department, 49–51, 53, 56, 66, 142
Woods, Huia, New Zealand officer, 5, 65
World War, First, 17, 24, 38–40, 101, 108
World War, Second, 3, 5–6, 43–5, 51, 53, 66, 69, 71, 76–7, 80, 106–7, 137

Xhosa, 11, 29–35, 107

Zaire, 104
Zambezi River, 33, 70, 72–3, 79, 84, 86–7
Zambia, 17, 40, 67, 70–3, 76, 79, 81, 86, 95, 97, 102–3, 105–6, 111, 131, 139
Zimbabwe, 6, 9, 11, 16, 21–2, 27, 33, 70, 75, 96, 131–2, 137, 142
Zimbabwe African Nationalist Liberation Army (ZANLA), 70, 73, 87–9, 91, 93, 95, 139
Zimbabwe African Nationalist Union (ZANU), 70, 73–5, 79, 84, 87, 96
Zimbabwe African National Union – Patriotic Front (ZANU–PF), 75
Zimbabwe African People's Union (ZAPU), 70–1, 73–5, 79, 87, 96, 102
Zimbabwe Commando Regiment, 96
Zimbabwe National Army (ZNA), 96
Zimbabwe Parachute Regiment, 96
Zimbabwe People's Revolutionary Army (ZIPRA), 70, 73, 89, 96
Zimbabwe Republic Police, 97
Zimbabwe Republic Police (ZRP), 96
Zulu, 11, 25, 27, 32